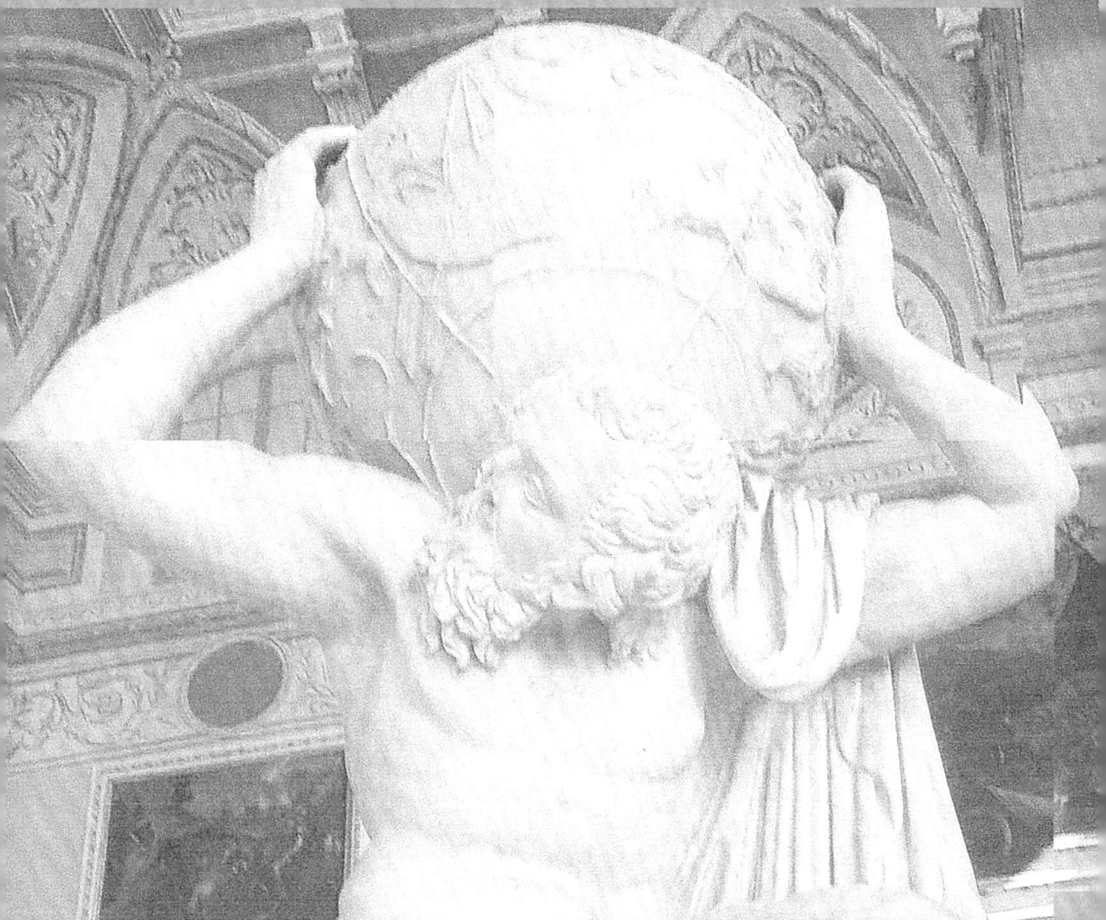

ART CONSERVATION

Jenny Stanford Series on Art Conservation — Volume 1

ART CONSERVATION

Mechanical Properties and Testing of Materials

Revised Edition

W. (Bill) Wei

JENNY STANFORD
PUBLISHING

Published by

Jenny Stanford Publishing Pte. Ltd.
Level 34, Centennial Tower
3 Temasek Avenue
Singapore 039190

Email: editorial@jennystanford.com
Web: www.jennystanford.com

British Library Cataloguing-in-Publication Data
A catalogue record for this book is available from the British Library.

Art Conservation: Mechanical Properties and Testing of Materials (Revised Edition)

ISBN 978-981-5129-63-2 (Hardcover)
ISBN 978-1-003-61644-3 (eBook)

Contents

Preface

Until a fateful day in 1992, my entire education and career were that of a typical materials engineer in the aircraft and power industries. On that day, my team-tennis doubles partner and good friend, Konrad Laudenbacher, who was a senior conservator at the Alte and Neue Pinakothek museums in Munich, Germany, thought that my background would serve the art conservation world much better. He invited me to visit his conservation studios at the Doerner Institut and arranged for me to do some volunteer work there.

I recall my first interview for that volunteer position with Dr. Andreas Burmester, head of the conservation science department, and later, head of the whole institute. At that time, the institute was involved in the European project, VASARI, looking at the use of the first high-resolution digital cameras for, among others, examining the condition of paintings before and after transport. When asked to compare two digital images of a painting before and after an international loan, I noted that a crack that appeared in the "before" image was closed in the "after" image. Dr. Burmester was surprised, exclaiming, "Cracks don't close!" I replied, "Of course they do. If the canvas bowed one way when the first image was taken, and happened to bow the other way in the image taken after transport, the crack would appear to close. That is the basis of cyclic loading and crack growth." I was eventually offered a volunteer position and worked there for a year and a half. This became my ticket to the world of conservation and conservation science.

What may seem like a nice personal story actually exposed a great gap in knowledge and expertise in the cultural heritage world which Dr. Burmester was trying to close and which still exists. When an object requires treatment, the initial technical research that is conducted is almost always chemical in nature. It is therefore not surprising, that the scientific curriculum of conservation training programs is strongly chemistry-oriented.

However, conservators face many questions in their daily work that involve the mechanical behavior of materials. Despite the number and complexity of such problems, mechanical properties and testing are almost never included as a regular part of any conservation training curriculum. It is, therefore, not surprising that there is much confusion and uncertainty in the conservation world when it comes to dealing with the mechanical aspects of restoration treatments.

I am proud that this book has been selected to be the first in the new *Jenny Stanford Series on Art Conservation*. It is an updated version of the original book, *Art Conservation: Mechanical Properties and Testing of Materials* published in 2021. The book aims to provide an initial remedy to this situation. The basics of the mechanical properties of materials and mechanical testing are introduced in a manner so that conservators without a background in engineering and physical sciences can understand their role in conservation treatments. The book is based in large part on lectures that I have given in one- or two-day introductory classes on the subject over the past decade in the conservation training programs at the University of Applied Sciences, Berlin, and the University of Amsterdam, as well as the research and consulting projects that I have conducted over the years. This updated version includes more examples and case studies, which are currently of interest to the conservation field, as well as a section on the planning of mechanical testing. In this way, I hope that the application of the basic theoretical concepts to daily practice will become even clearer.

On that note, I would like to thank Konrad, for encouraging me to enter the world of art conservation, and Dr. Burmester, for giving me my first access to that world at the Doerner Institut. I would also like to thank my colleague IJsbrand Hummelen for welcoming me to what now is the Cultural Heritage Laboratory of the Cultural Heritage Agency of the Netherlands (RCE) and his critical reading of this manuscript. Furthermore, I appreciate the time and effort that my colleague, private conservator Rebecca Rushfeld, Queens, NY, put in to make sure that the book is readable for professionals without a technical/engineering background. My deepest thanks also go to my international colleagues in conservation science—Susan Barger, conservation scientist and consultant for small museums and archives, and Jerry Podany, former head of antiquities conservation at the J. Paul Getty Museum—for their critical technical review of the book. My thanks also go to the director of the RCE, Susan Lammers, and the head of the Cultural Heritage Laboratory, Bauke Zeilstra, for providing me with the time and support to write the original version of this book. Finally, I would like to thank my wife, Ida van der Lee, a community/ritual artist, for her love and support during this project.

W. (Bill) Wei
Wilbar Holding B.V. / vibmech.nl
Amsterdam, The Netherlands
November 2024

Chapter 1

Introduction

The scientific and technical world of the conservation and restoration of art and other objects of cultural heritage is largely based in chemistry. This is evident in the majority of questions which must be answered when making conservation decisions, such as

- what is the chemical composition of the materials in an object,
- what is the state of aging of the object,
- what kinds of chemicals and materials can be safely used for a conservation treatment, and
- what climate conditions, *e.g.*, light, temperature and relative humidity, can an object be displayed and stored in?

However, conservators face a large number of questions in their work which involve the mechanical behavior of materials. Examples of problems which they have to deal with include

- bending and cracking in wooden objects such as panel paintings and furniture,
- reinforcing or repairing joints in statues,
- hanging of textile wall tapestries (Fig. 1.1a),
- craquelure in oil paintings (Fig. 1.1b),

Art Conservation: Mechanical Properties and Testing of Materials
W. (Bill) Wei

ISBN 978-981-5129-63-2 (Hardcover), 978-1-003-61644-3 (eBook)
www.jennystanford.com

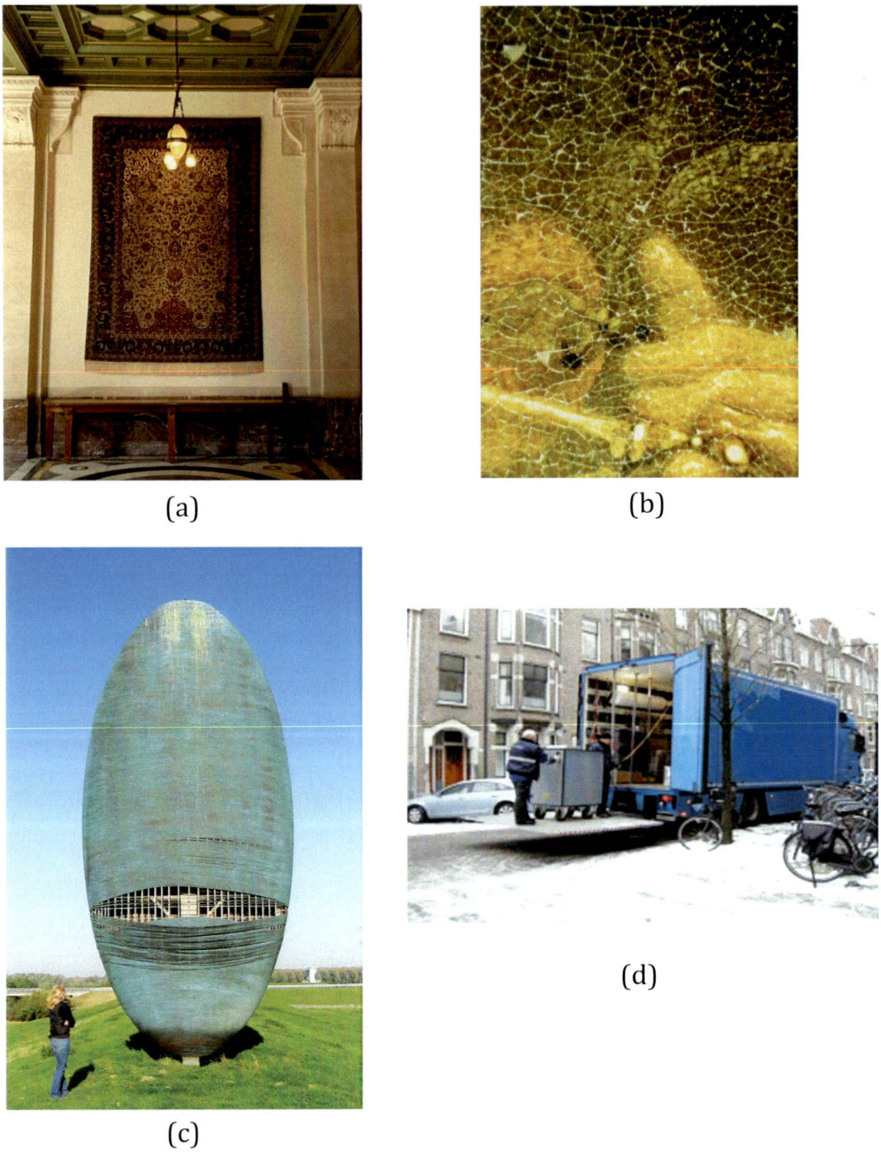

Figure 1.1 Examples of mechanical issues which face conservators: (a) Hanging of a wall tapestry, (b) craquelure in an oil painting, (c) failure of bonds holding copper coils to the metal frame of an outdoor sculpture ("Tong van Lucifer" (1993) by Ruud van de Wint, located on the Knardijk near the A6 highway, The Netherlands), and (d) vibration and shock during the transport of works of art.

- deformation and wear of metal objects,
- effects of (cyclic) changes in temperature and relative humidity on (outdoor) sculptures (Fig. 1.1c),
- paint loss,
- repair of tears and cuts in canvas paintings,
- strength of adhesives and consolidants,
- tension in painting canvasses, and
- vibrations and shock during the transport of objects, or during construction and other events in and around museums (Fig. 1.1d).

There are, however, other situations where it is not as obvious that mechanical stresses play an important role in the long-term condition of objects. Many such situations are related to the climate conditions noted above, including

- the effects of cyclic changes in temperature and relative humidity on canvasses,
- the effects of cyclic changes in temperature and relative humidity on wood panel paintings or wooden furniture,
- cracking of reinforced concrete art and architecture due to corrosion of the steel reinforcement bars,
- the concept of internal stresses, and
- dimensional changes (expansion or shrinkage) of adhesives.

The scientific curricula of virtually all conservation training programs around the world concentrate on the chemistry and physical properties of materials. Mechanical properties are almost never taught to any extent. It is therefore not surprising that there is much confusion, misinformation and uncertainty in the conservation world when it comes to dealing with mechanical aspects of restoration treatments.

In spite of this, when asked what terms they can think of that have to do with mechanical properties, most conservation students surprise themselves with how many words they can come up with. An alphabetical list of words compiled by the author in twelve years of teaching daylong workshops on mechanical properties is shown in Table 1.1. If one now reorganizes these words into related groups, one sees that there are four main categories related in some way to the words “force,” “weight,” “deformation” and “failure,” (see Table 1.2). These terms lie at the core of the engineering concepts of the mechanical properties of materials in this book.

Table 1.1 Words related to mechanical properties familiar to conservation students

Break	Expansion (thermal)	Strength
Brittle	Force	Stress
Contraction	Friction	Stretch
Crack	Gravity	Swelling
Damping	Kilogram	Tear
Deformation	Loose	Tension
Density	Mass	Tight
Dent	Plastic	Wear
Elastic	Pressure	Weight
Elongation	Shock	
Energy	Shrinkage	

Table 1.2 Categorization of the mechanical property terms in Table 1.1

Force	Weight	Deformation	Failure
Damping	Density	Brittle	Break
Energy	Kilogram	Contraction	Crack
Force	Mass	Deformation	Tear
Friction	Weight	Dent	Wear
Gravity		Elastic	
Loose		Elongation	
Pressure		Expansion (thermal)	
Shock		Plastic	
Strength		Shrinkage	
Stress		Stretch	
Tension		Swelling	
Tight			

The objective of this book is to teach the basics of the mechanical properties and testing of materials. While learning these basics, it is, however, important to remember that every profession, be it art conservation, chemistry, history or engineering, has its own language and vocabulary. As with any foreign language, certain words mean certain things. They may look like something in another language, but if they are used in the wrong way, misunderstandings and confusion will certainly be a result. This should be taken into consideration when an engineer with no background in art or conservation is consulted on some aspect of mechanical testing.

This book begins with two chapters on the most important concepts and terminology of the mechanical properties of materials. Key terms are presented in bold-faced type the first time they are introduced. The most commonly used methods for mechanical testing are then described in the fourth chapter. The book concludes with a discussion of a number of advanced concepts which conservators may come across in their work. It is hoped that the minimum which the reader comes away with after reading this book is the proper use of the terminology and language of this subject.

It is impossible to completely avoid mathematics in a field such as this, but the mathematics in this book has been limited to basic equations for the understanding and calculation of the mechanical properties of materials. Furthermore, all units for the various quantities given in this book are expressed in standard international (SI) units, in line with standard international scientific practice. Appendix 1 provides conversions between the SI units used in this book and their English (Imperial) counterparts which are still commonly used in industry, among others, in the United States of America.

Before continuing, two words of caution. First, for many situations, the basic concepts and exercises presented in this book may be sufficient for providing a rough answer to a mechanical issue during the treatment of an object. However, for most of the unique and/or complex combinations of materials and forms found in works of art and objects of cultural heritage that a conservator will encounter, more care must be taken in determining testing requirements, and outside expertise may be required. The concepts presented in this book are meant to help conservators and other museum professionals understand what is required in approaching the question. Second, a number of materials properties are presented in this book in order to illustrate certain concepts or for use in the exercises. While these are real values, they represent general values for the materials. They are not meant to be used for specific situations. Readers should consult the proper literature, experts and suppliers to determine their specific needs.

Chapter 2

Static Mechanical Properties

As the reader may have noted in Chapter 1, there are two main types of mechanical loads which conservators will be confronted with in their work:

- static, where loads and the resulting deformation do not change with time, and
- dynamic, where loads and the resulting deformation do change with time.

In this chapter, we will look at the basic concepts and terminology for static loads. What is meant by a static load is a load which is applied slowly and in one motion, such as sitting down on a chair or placing a flower pot on a table, and then does not change with time. In Chapter 3, we will examine the basic concepts and terminology for dynamic loads.

Looking back at the vocabulary list of mechanical properties in Tables 1.1 and 1.2 in Chapter 1, it can be seen that many of the words used in daily life have to do with the response of materials to static loads, that is, their static mechanical properties. However, in lay terms, many of these words can mean different things in different contexts. This has often led to confusion and misinterpretations of tests published in the conservation literature. In the following sections, we

Art Conservation: Mechanical Properties and Testing of Materials
W. (Bill) Wei

ISBN 978-981-5129-63-2 (Hardcover), 978-1-003-61644-3 (eBook)
www.jennystanford.com

will reduce these words to the basic technical terminology for static mechanical loads used in science and engineering. This will help with the proper design and analysis of tests for determining the mechanical properties of materials which may need to be used in a conservation treatment. It will also help in communications with engineering professionals who may be consulted.

2.1 Force vs. Stress

Imagine that a testing engineer is given two rods as shown in Fig. 2.1. They have the same length, l_0, and are made of exactly the same material, for example, the same steel alloy with exactly the same chemical composition and microstructure. The only difference is that rod 1 is thicker. The engineer is asked to test the strength of the rods by pulling them parallel to their length, that is, in **tension**. Which rod is stronger?

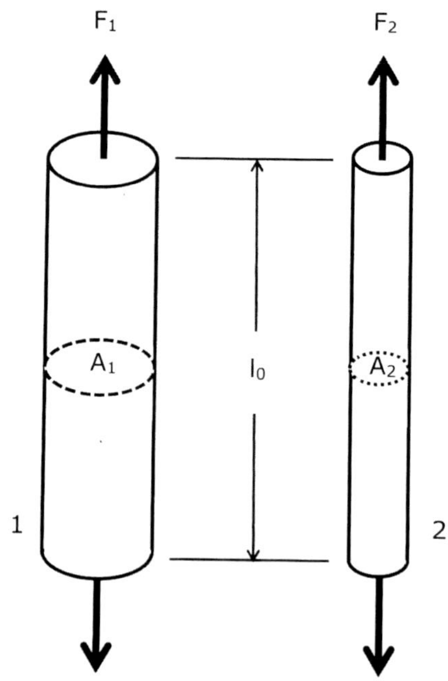

Figure 2.1 Two rods of the same length and exactly the same material. Which is stronger?

Some readers might choose rod number 1 because it is thicker. Others would say that they are the same because the materials are exactly the same. If the reader were to do the test, he/she would get a result which would support the first answer based on common sense. Common sense would say that one would need more force, F_1, to break rod 1 than one would need, F_2, to break rod 2 because rod 1 is thicker. But both rods are made of exactly the same material, so shouldn't they have the same strength? And what is the value of that force? To do this simple test, one would probably hang weights on the rods, in say, kilograms.

The answer is that the material can have only a single value of strength. Of course, one needs more force to break rod 1 because it is thicker than rod 2. The fact that rod 1 is thicker than rod 2 is, however, a difference in a geometric property of the two rods. The thickness of a rod is not a materials property.

However, if we divide the force required to break the two rods by the cross-section area carrying those forces, we will find that we will get the same answer, that is, force 1, F_1, divided by the cross-section area of rod 1, A_1, will give us the same answer as force 2, F_2, divided by the cross-section area of rod 2, A_2, or in equation form,

$$\frac{F_1}{A_1} = \frac{F_2}{A_2} \qquad (2.1)$$

The force divided by the cross-section area gives us what is known as a value of **stress**, and as can be seen, the stress when the two rods break is the same in both cases. Therefore, if we express the **strength** of a material using this term, stress, we will get the same answer for both rods of exactly the same material, as should be expected.

Let us now continue this discussion with some key definitions. In physics and engineering, **force** is defined based on Newton's laws of motion as an interaction with an object which will change its motion. However, if the object is fixed in place, as in most of the cases you will face as a conservator, a force is that which will deform it. The unit for force is defined mathematically based on the Newtonian force as **mass** times acceleration, that is,

$$F = m \times a, \tag{2.2}$$

where m is mass, the amount of a material in kilograms (kg), and a is acceleration in meters/second^2 (m/s^2). As might be suspected, what is being hung on the rods to do the test are not forces, but masses in kilograms. The unit of force is called a Newton (N) and is defined as

$$F = m \times g, \tag{2.3}$$

where g is the acceleration due to gravity on Earth, that is,

$$g = 9.8 \text{ m/s}^2.$$

What was hung on the rods in the test in Fig. 2.1 was not a weight (the lay term), but a mass. Technically speaking, **weight** is the force in Newtons which the mass exerts on the rods, or as another example, which you exert on the ground. Weight is thus also mass × g. A person weighing 60 kg exerts a force on the ground of

$$F_{\text{earth}} = 60 \text{ kg} \times 9.8 \text{ m/s}^2 = 588 \text{ N}.$$

Note that we exert less force on the ground on the moon because the acceleration due to gravity there is only 1.6 m/s^2. The force our person exerts on the moon is then

$$F_{\text{moon}} = 60 \text{ kg} \times 1.6 \text{ m/s}^2 = 96 \text{ N}.$$

Mass is thus the same everywhere in the universe. However, weight, that is, the force exerted on the ground, or that is exerted on the rods in Fig. 2.1, depends on the acceleration of gravity.

Forces can be **tensile**, pulling the rods in Fig. 2.1 apart. Forces can also be **compressive**, that is, in the opposite direction of the arrows in Fig. 2.1, crushing the rod, or a person's weight pushing on the ground or the floor they are standing on. Examples of tensile forces include the stretching of painting canvasses (Fig. 2.2a), the forces in the shoulders of a historic garment hung on a hanger or mannequin (Fig. 2.2b), or the forces in the chain holding a chandelier (Fig. 2.2c). Examples of compressive forces include those on the feet of lead organ pipes which have to carry the weight of the whole pipe (Fig. 2.2d), or in the legs of a table due to the weight of the table top and anything on the table (Fig. 2.2e). Objects can also have combined tensile and compressive loading if they are loaded in a bending mode, for example, in the shoulder area of a statue (Fig. 2.2f).

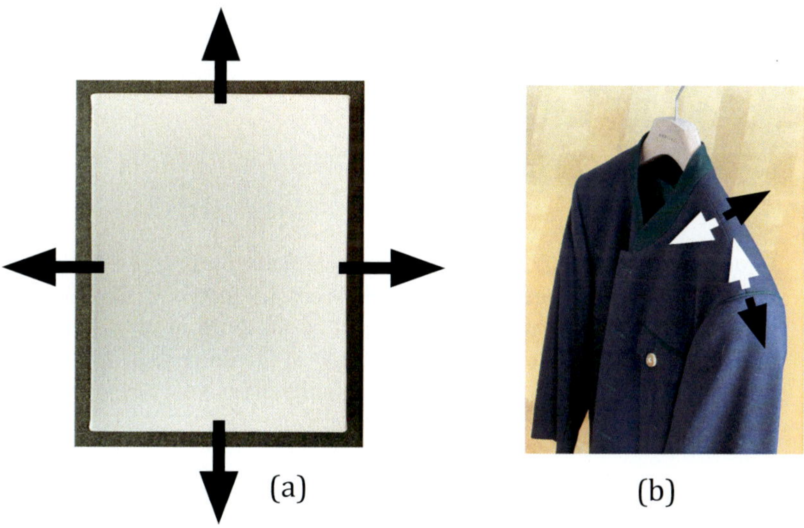

Figure 2.2 Examples of tensile and compressive forces in cultural heritage objects: (a) Tensile forces ⇔ (in both the vertical and horizontal directions) used to stretch a painting canvas. (b) Tensile forces ⇔ caused by the weight of an historic garment on a clothes hanger.

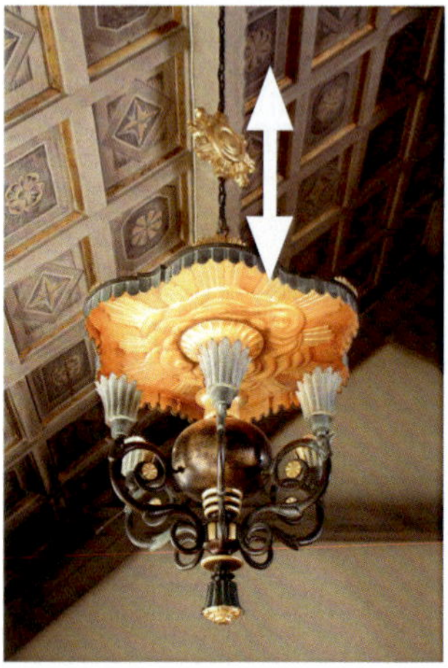

(c)

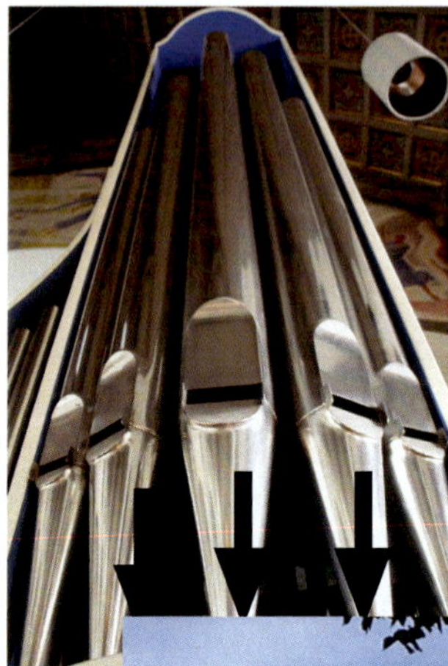

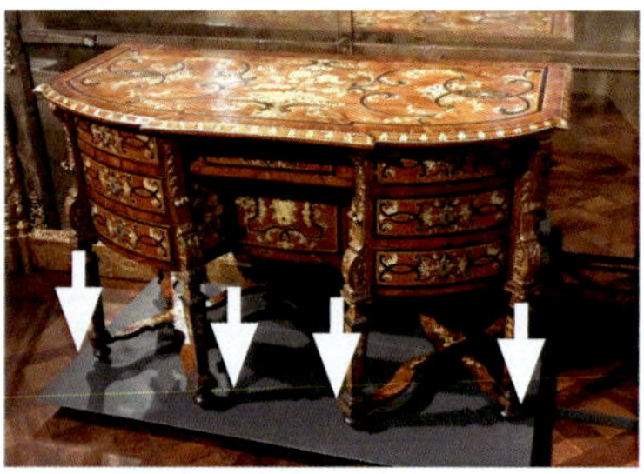

(e)

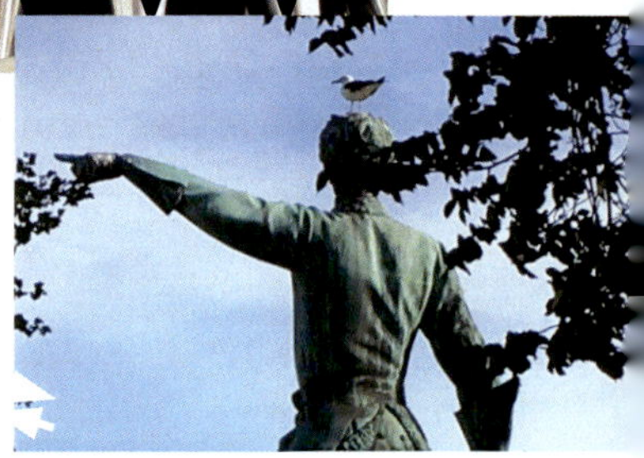

(f)

Figure 2.2 (*continued*) Examples of tensile and compressive forces in cultural heritage objects. (c) Tensile forces ⇔ in the chain carrying a chandelier. (d) Compressive forces ⇓ at the bottom of a 4 m-long lead organ pipe. (e) Compressive forces ⇓ in furniture legs and on the podium. (f) Combined tensile ⇔ and compressive ⇒⇐ forces of the shoulder area of a statue in bending.

Forces can also be applied parallel to the load-carrying area as shown schematically in Fig. 2.3. Such forces are known as **shear** forces. Rubbing one's hands to keep them warm is an example of a shear force. Examples of shear loading in cultural heritage include the load put on Velcro® strips when they are used to hang tapestries (Fig. 2.4) or the load on hinges used to mount photographs or prints.

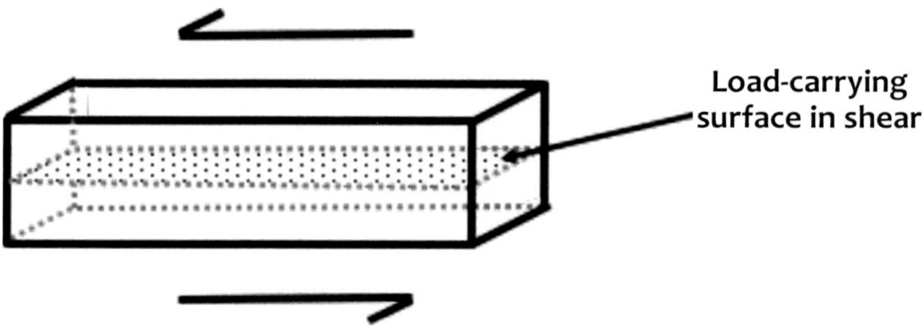

Figure 2.3 Shear loading where the force is applied parallel to the load-carrying area.

As we have seen, the force necessary to break the specimen in Fig. 2.1 depends on the geometry of the object. In order to be able to compare materials properties we need to transform the force into a unit that is independent of geometry. That is the unit which was introduced, stress. As we saw above with respect to Fig. 2.1, stress is defined mathematically as force, *F*, divided by the area carrying it, *A*, that is,

$$\sigma = \frac{F}{A} \tag{2.4}$$

where stress is written mathematically with the small Greek letter sigma, σ. The unit of stress is Newton per square millimeter, that is, N/mm^2. Note that one will often see stress given in the literature and on the Internet in units of mega-Pascal (MPa, note the double capital letters). Fortunately, the conversion is simple: 1 MPa = 1 N/mm^2, that is, the two units are synonymous.

For shear stresses, the small Greek letter tau, τ, is used instead of σ, that is,

$$\tau = \frac{F}{A} \tag{2.5}$$

where the force is now parallel to the load-carrying surface (see Fig. 2.3).

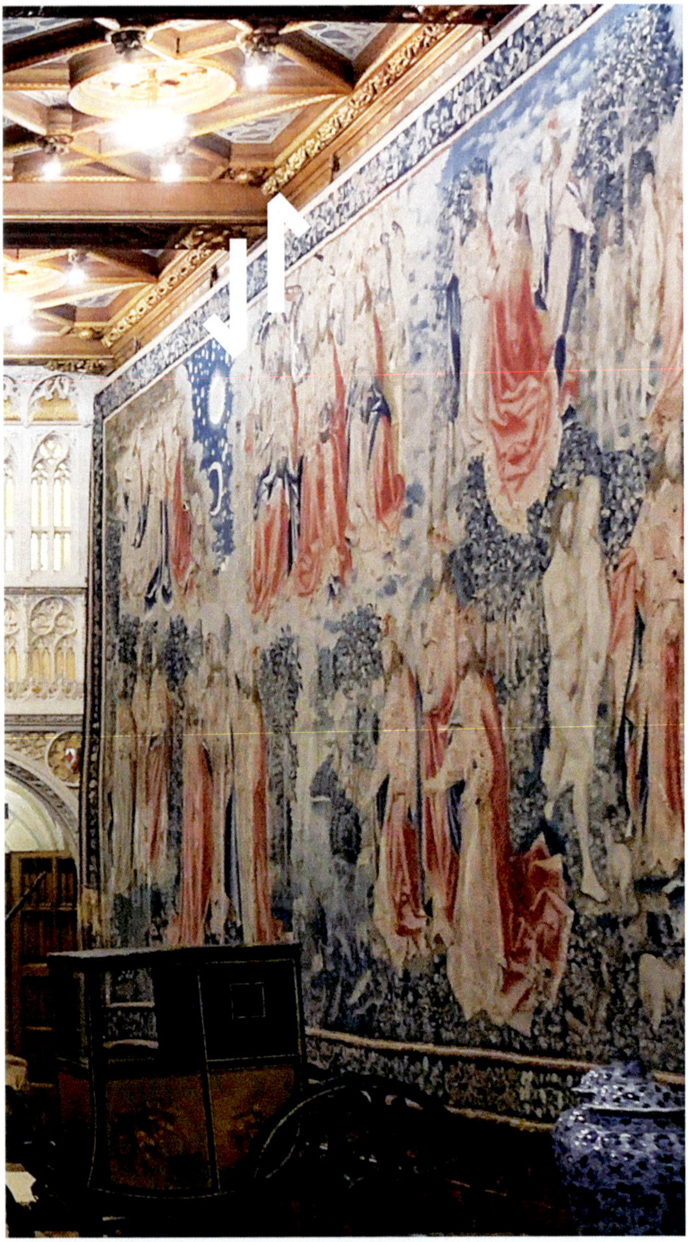

Figure 2.4 Example of shear forces (white arrows at top of photo) in a cultural heritage object: shear forces on Velcro® strips caused by the weight of a wall tapestry.

The **strength** of a material is defined as the stress at which it fails, in lay terms, when it breaks. The strength of a material is thus given in terms of the word stress. If one looks for materials' strengths from a supplier, one will always get information and tables in units of stress, N/mm^2 or MPa. If one looks at the packages of many commercial glues and adhesives, the strength of the glue will always be given in terms of shear stress. One might think practically that this is very inconvenient. Why can't the suppliers just give a straight answer and say which material will handle the forces on my object?

One must think of the answer from the viewpoint of the materials supplier. Conservators will be working with many different types and sizes of objects. They may, for example, have to choose between different types of glues for a broken 18th century wooden chair leg, or a slender, broken wooden leg of a 20th century Danish designer sofa. In the industrial world itself, engineers design a huge variety of different products and must choose among hundreds, if not, thousands of materials. It is impossible for materials or glues/adhesives manufacturers to provide information to compare materials if they have to test and provide catalogue information for every single geometric possibility and combination of forces that could come up. They thus provide standardized (or in engineering terms, normalized) test data on the material strength based on a standard cross-section area, that is, mm^2, and thus on stress, in N/mm^2 or MPa.

The consequences of this for conservators, and for engineers and designers in general, are as follows:

- If one is dealing with a broken part of an object and have decided to repair it in some way such as replacing it or gluing it, one first needs to look at the practical situation in terms of forces, F (or in another lay term, loads). What force/load does the repaired part or the repair material or glue need to carry?
- One then needs to translate that force/load into terms of stress, σ or τ, by dividing the required or desired load/force by the area carrying the force/load.

- The manufacturer or the literature can then be consulted in order to compare materials and select those with sufficient strength in terms of stress.

The terms we have just discussed are summarized in Table 2.1.

Table 2.1 Terminology related to force vs. stress

Term	Definition/calculation	
Acceleration – in general (a) – due to gravity (g)	$g = 9.8\ \text{m/s}^2$ (on Earth)	
Area, load-carrying (A)	Circle Rectangle	$A = \pi r^2$ ($\pi = 3.14$; r = radius) A = length × width
Force (Newtonian, F)	$F = m \times a$	(Eq. 2.2)
Force due to gravity (weight)	Weight = $m \times g$	(Eq. 2.3)
Stress (σ)	$\sigma = F/A$ (tensile, compressive) $\tau = F/A$ (shear)	(Eq. 2.4) (Eq. 2.5)

Practical exercise 2.1

A conservator is working on a wooden object and there is a joint which needs to be repaired as shown in Fig. 2.5. It is a glued connection which has to carry a mass of 10 kg. There are four kinds of glue in the conservator's studio with the shear strengths given in Table 2.2. The shear strength of the wood itself is 3.0 MPa. Which glue would the conservator select?

Table 2.2 Strengths of four kinds of glue for practical exercise 2.1

Glue	Shear strength (τ) (MPa)
A	2.0
B	2.6
C	3.2
D	4.0

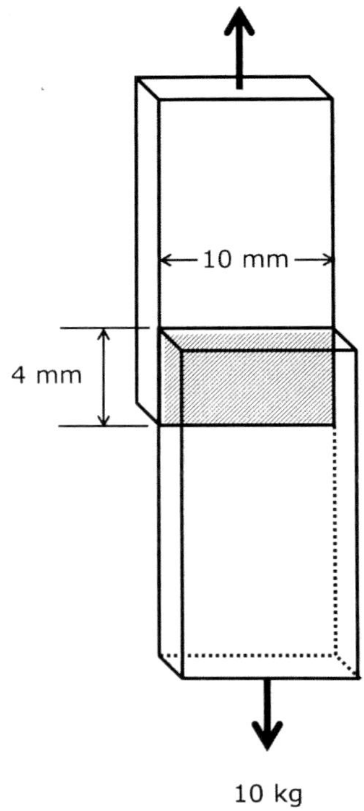

Figure 2.5 Wooden connection which needs to be glued.

Solution

The glue joint is loaded in shear. The force on the load-carrying area (hatched area in Fig. 2.5) is, using Eq. 2.3,

$$F = \text{mg} = 10 \text{ kg} \times 9.8 \text{ m/s}^2 = 98 \text{ N}.$$

The load carrying area is

$$A = 4 \text{ mm} \times 10 \text{ mm} = 40 \text{ mm}^2.$$

The shear stress on the glue joint is then, using Eq. 2.5,

$$\tau = \frac{F}{A} = \frac{98 \text{ N}}{40 \text{ mm}^2} = 2.45 \text{ N/mm}^2 \text{ or } 2.45 \text{ MPa.}$$

Comparing this value with the shear strengths of the glues in Table 2.2, one can see that glues B, C, and D have sufficient strength to carry the 10 kg.

However, if this cultural heritage object is being restored for use, for example, in an historic house, it does run the risk of breaking again at this joint. One must then think about more than just the shear strength at the joint. The wood itself has a shear strength of 3.0 MPa. If one were to use glues C or D, one would be using a glue which is stronger than the wood under shear loading at the joint. That means that if the joint fails, it will not be the glue that fails this time, but the wood. To avoid that risk, glue B would have to be selected, which has sufficient strength to carry the 10 kg. With glue B, if the joint fails, it will be the glue joint that fails and not, say, the splintering of the wood right next to the joint.

Practical exercise 2.2

Imagine that a beginning conservator can only afford glue A, but needs to repair the joint in Fig. 2.5 with the same loading conditions as given in Practical exercise 2.1. However, the client does not require the conservator to hold strictly to conservation ethics. What could be done in this case?

Solution

One could increase the load-carrying area, that is, increase the area that is glued to reduce the stress in the joint. In this case, the force is known, 98 N, and the strength of glue A is 2.0 MPa or 2.0 N/mm^2. One can then calculate the minimum area which would be needed to reduce the stress by rewriting Eq. 2.5 for (shear) stress,

$\tau = F/A$, as

$$A = \frac{F}{t} = \frac{98\ \text{N}}{2.0\ \text{N/mm}^2} = 49\ \text{mm}^2.$$

2.2 Elongation/Deformation vs. Strain

Imagine now that the engineer is given two rods as shown in Fig. 2.6. Again, they are made of exactly the same material, for example, the same steel alloy with exactly the same chemical composition and microstructure. They also have the same cross-section area, *A*. However, rod 1 is shorter than rod 2. Suppose they are loaded with the same force, *F*, and thus the same stress, σ. Which rod will become longer?

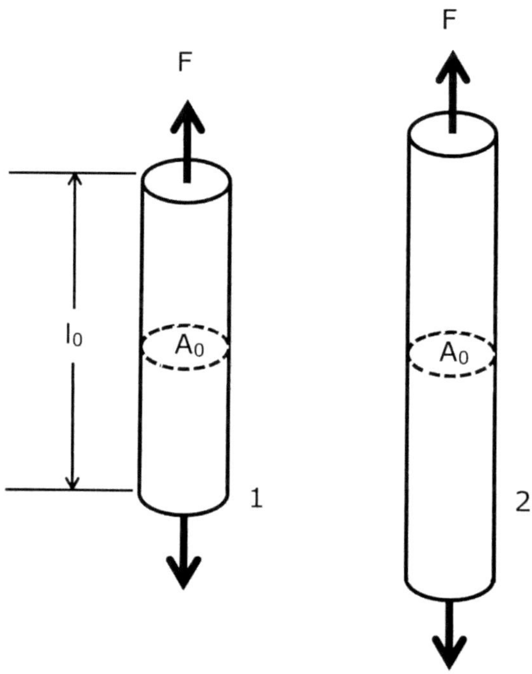

Figure 2.6 Two rods of exactly the same material undergo the same stress. Which one will become longer?

If one selected rod 2 one would be correct in the geometric sense. The longer rod 2 will elongate more in, for example, millimeters, simply because it has more material than rod 1. However, both rods are made of exactly the same material, and the stress is the same. We have already seen that geometry is not a materials property, so again, it should seem odd that one rod deforms more than the other. However, if we perform a similar "normalizing" calculation as we did for force and stress, we will find that the fractional change in length will be the same for both rods.

The fraction change in length is known as **strain**. Mathematically it is the change in length of the rod, $l - l_o$ divided by the original length l_o, that is,

$$\varepsilon = \frac{(l - l_o)}{l_o} \tag{2.6}$$

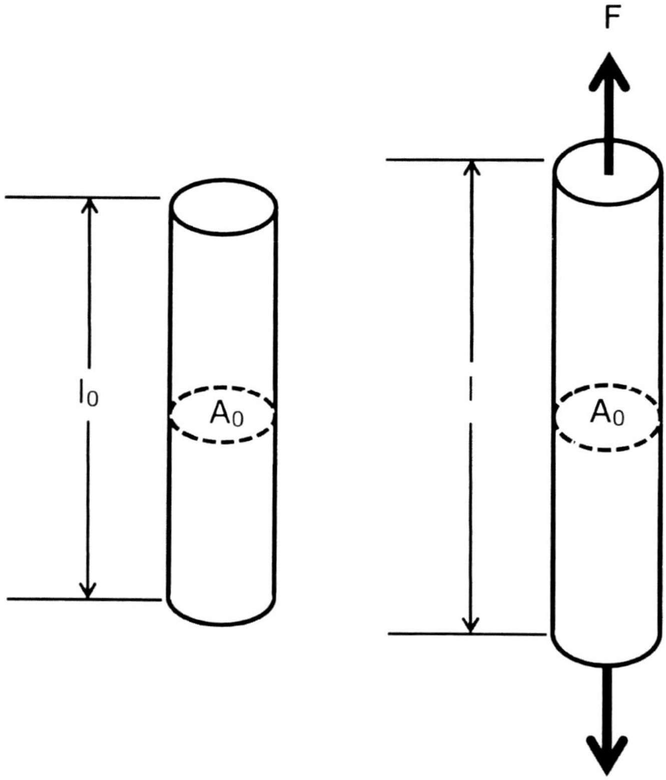

Figure 2.7 Change in length of a rod due to a stress, F/A_0.

where the various dimensions are defined in Fig. 2.7, and ε is the Greek letter for strain. $l - l_0$ is often abbreviated as Δl, where the Greek letter Δ is used in mathematics to mean "change," that is,

$$\varepsilon = \frac{\Delta l}{l_0}. \tag{2.7}$$

Note that the dimensions of strain are length divided by length, so that strain is a dimensionless fraction. This may not be so intuitive for some readers, so it may be easier to express strain as a percentage change in length. This can be done by multiplying the fraction, ε, by 100, to get the percentage change in length, that is,

$$\varepsilon(\text{in percent}) = 100 \times \frac{(l - l_0)}{l_0}. \tag{2.8}$$

In conservation, we are generally dealing with stresses and eventually things breaking when the stress reaches the material's strength. However, in many engineering/industrial applications, and in some exceptional cases in conservation, especially of industrial heritage objects, one failure criterion is that something deforms too much. The next time the reader flies in a passenger airplane, take a look at the front of the jet engine. There, one will see the so-called fan turning. In operation, each fan blade will be loaded by a centrifugal force, which translates into a stress, which further translates into strain of the blade during flight. It is not particularly pleasant if the strain is so great that the fan blade comes in contact with the turbine engine housing. Thus, analogous to the force/stress situation, engineers who design these parts cannot go to suppliers and expect them to tell them how much a material will lengthen under a given force. They must translate that into strain under a given stress. The supplier can then provide information on the strain behavior of materials, information which is useful for every client.

2.3 The Relationship between Stress and Strain

Stress and strain are related as one might expect. As the force is increased for the tensile test discussed in Fig. 2.1 or 2.6, a graph will be obtained showing the relationship between the applied force on the specimen and the change in length for a typical metal (see Fig. 2.8a). These results can be calculated into stress and strain based on the original load-carrying area of the specimen, A_0, and the original length of the specimen l_0, as defined in Fig. 2.6. This results in a so-called engineering stress–strain (σ–ε) diagram as shown schematically in Fig. 2.8b.

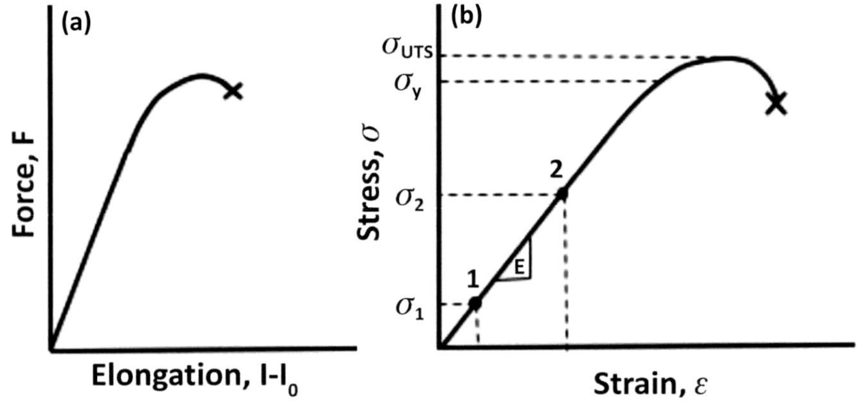

Figure 2.8 Schematic (a) load elongation and (b) stress–strain diagrams for a typical metal.

For most metals, the increase in strain with an increase in stress begins as a straight line. This is called "linear" behavior. This part of the graph also describes something known as elastic behavior. If you stressed the material to, say point 2 in Fig. 2.8b and then unstressed it, the specimen would go back to its original length. A material is thus called **elastic** in technical terms when it returns back to its original form after being deformed. If the stress–strain curve is a straight line as in the lower part of Fig. 2.8b, we say that the material shows **linear-elastic behavior**.

At a certain point for metals, the stress level, σ_y, the curve starts to bend and eventually reaches a maximum at the stress level, σ_{UTS} in the tensile case, or more generally, σ_f. These three stress parameters will be discussed in the following section of this chapter. The curve then bends down to the X which marks the point at which the specimen breaks or technically, **fractures**. It could seem odd that the stress would decrease before fracture, but this seemingly odd part of the curve will also be discussed in the following section of this chapter.

The slope (steepness) of the straight part of the graph in Fig. 2.8b is an important materials property known as the **elastic modulus**, *E*, or *E*-modulus. In the engineering world it is also known as **Young's modulus**, named after the British scientist who first described its calculation. The *E*-modulus is calculated by dividing the change in stress between, say, points 1 and 2 in Fig. 2.8b by the change in strain between those points, in equation form,

$$E = \frac{(\sigma_2 - \sigma_1)}{(\varepsilon_2 - \varepsilon_1)} \tag{2.9}$$

To make things a little simpler, if we were to choose point 1 to be the beginning of the stress–strain curve (lower left-hand point of the diagram in Fig. 2.8b) so that σ_1 and ε_1 are both zero, we would get

$$E = \frac{\sigma_2 - 0}{\varepsilon_2 - 0}$$

or more generally

$E = \sigma/\varepsilon$, which can also be written as

$$\sigma = E \times \varepsilon \tag{2.10}$$

This equation is known as **Hooke's law**, and we will discuss this in more detail in the next section.

The elastic modulus describes the **stiffness of a material**. It tells us how much stress we need to strain the material a certain amount. Most ceramic materials are stiff in this sense (see Fig. 2.9). Many ceramics are very strong, but do not deform very much. They therefore have a very steep linear elastic region as compared with metals, meaning that they have very high values of *E*-modulus. On the other end of the scale in Fig. 2.9, elastomers such as rubber have much lower strength than metals and ceramics, but can be strained to over five times their length before they break. The value of the *E*-modulus for elastomers is thus much lower than for ceramics and metals. Note that many plastics and textiles show elastic behavior, but it is not linear; see schematic stress–strain curves in Fig. 2.10. Some textiles may also have the jagged failure curve found for wood.

In terms of terminology usage, beware that the material stiffness (*E*-modulus) is not the same as the geometric stiffness of an object. The use of a material with a high *E*-modulus, that is, a stiff material, does not guarantee that a structure will be stiff. On the other hand, many sufficiently stiff structures can be made with materials with lower stiffness (lower *E*-modulus). For example, a residential home often has a wooden frame rather than a steel one.

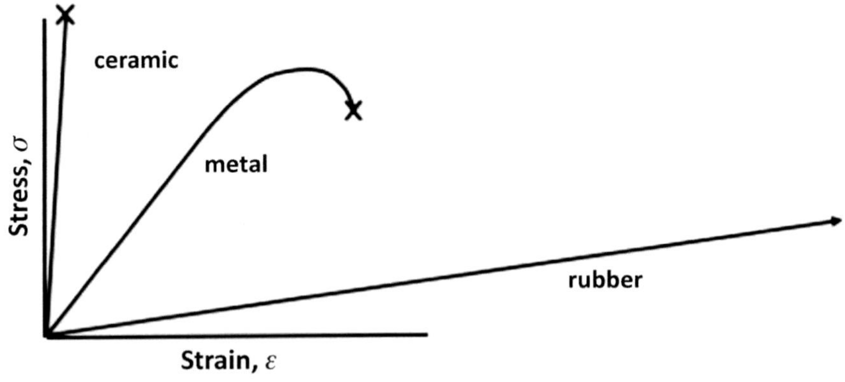

Figure 2.9 Typical (schematic) stress–strain diagrams for ceramics, metals and elastomers such as rubber. Note that the schematic curve for metals is the same as in Fig. 2.8.

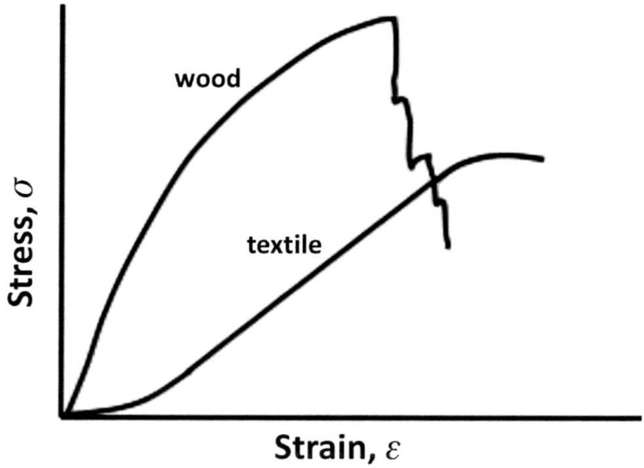

Figure 2.10 Schematic stress–strain diagrams for a textile and for wood (in bending).

2.4 Elastic vs. Plastic Deformation

We have seen that as a material is stressed, it deforms to a certain amount of strain. One can also say this the other way around. As a material is strained, it develops a certain amount of stress. In the previous section, we briefly introduced the concept of elastic behavior. A material behaves elastically if, after being strained or stressed, it returns to its original form. The strain is recoverable, and is known as **elastic deformation**. For many materials, this occurs linearly as we saw in Fig. 2.8b, and is described by Eq. 2.9, Hooke's law. A material will follow the same straight line as it is stressed/loaded and unstressed/unloaded. You can think of the material as a spring (see Fig. 2.11). When you exert a force, you stretch the spring. When you reduce the force back to zero, the spring returns back to its original form. Note that in basic physics, the relationship between the force and change in length of a spring looks like Eqs. 2.9 and 2.10. Instead of the E-modulus, one speaks of a spring constant. In fact, the atomic bonds in a material can be considered to be springs. When you stress the material within the linear-elastic range, the bonds will stretch, that is, the distance between the atoms will increase. When you unload the material, the bonds will return to their original length and unstressed position.

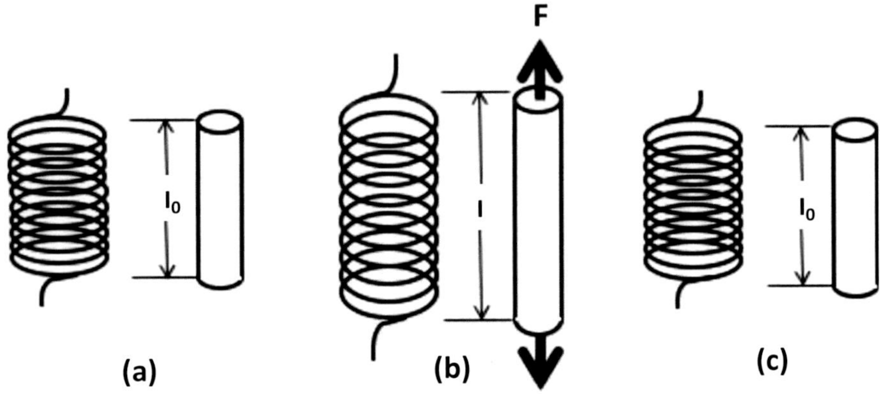

Figure 2.11 The linear-elastic behavior of materials (Figs. 2.7 and 2.8) is equivalent to the deformation of a spring. When a spring (a) is loaded it stretches (b) and when it is unloaded it returns to its original length (c).

There are, however, many materials, in particular, polymers and plastics, which show non-linear elastic behavior as shown schematically in Fig. 2.12. When loaded and unloaded, the strain will return to zero, but the stress–strain diagram in this region is not a straight line.

As the stress increases, the material will reach a state where it begins to permanently deform. This permanent deformation is known as **plastic deformation**. If the material is unloaded, it will not return to its original form. There will be some permanent deformation. For a metal, plastic deformation begins at the so-called yield stress, σ_y (see Fig. 2.13 which is a simplified version of Fig. 2.8b where this parameter was first introduced). If the material is unloaded from this stress level, it will unload along the dot-dashed line parallel to the linear elastic part of the diagram, removing the elastic strain it has built up. However, there will be some deformation remaining. Referring again to the analogy of the spring in Fig. 2.11, if the spring is overloaded, it will only return partially (the elastic part). (Note that the analogy of the atomic bond does not hold here. The explanation for this can be found in most materials science textbooks and is beyond the scope of this book.)

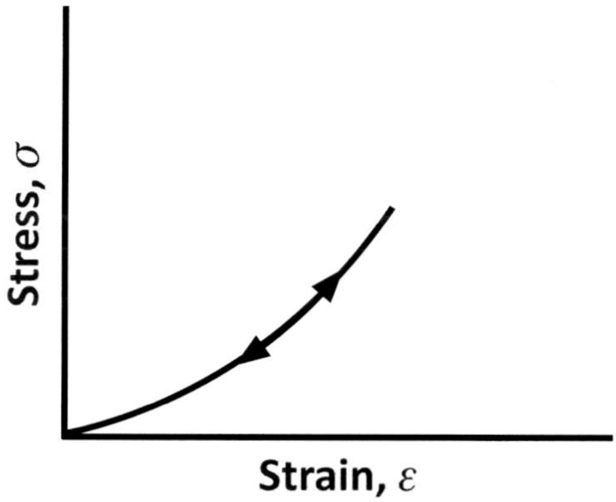

Figure 2.12 Elastic part of a stress–strain diagram for a material that shows non-linear elastic behavior.

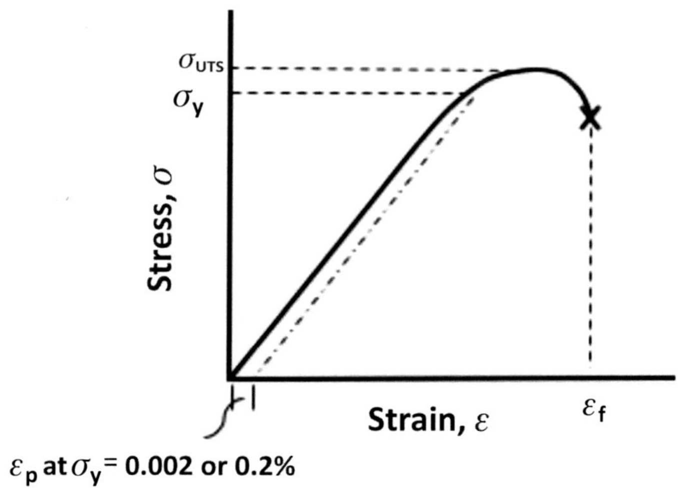

Figure 2.13 Definition of the yield stress, σ_y and plastic deformation, ε_p. Note that this diagram is a simplified version of Fig. 2.8b.

Notice that the stress–strain curve for a metal does not actually have a clear point which can be defined as the beginning of plastic deformation and the yield stress, σ_y. International agreement has set σ_y as the stress at which 0.002 or 0.2% permanent strain remains when the specimen is unloaded. One will therefore often see the symbol $\sigma_{0.2}$ instead of σ_y.

As the stress increases, the amount of plastic deformation will increase until the stress reaches a maximum at a level known as the "ultimate tensile strength," σ_{UTS}, or the failure or fracture stress, σ_f. σ_{UTS}, or σ_f, is none other than the **material strength**. The corresponding strain at failure is known as the fracture strain, ε_f, see Fig. 2.13.

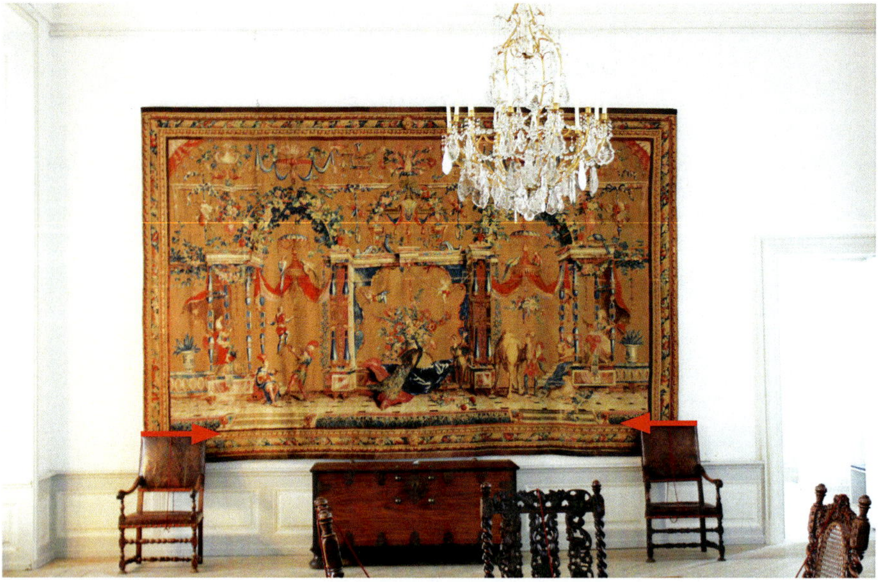

Figure 2.14 Example of the Poisson effect, a slight wrinkling of the bottom edge of a wall tapestry (red arrows) supported only at the top edge.

As we noted in the previous section, the curve actually bends downwards to a lower stress level before reaching the fracture strain. This odd bend is a result of the fact that in engineering, the original area of the specimen, A_0, is used to calculate the stress using Eq. 2.4, no matter what the force is. However, if a rod is loaded with a tensile force as shown in the schematic drawings, the rod will have to get thinner as the force increases. This is known as the **Poisson effect**, and is purely the result of the concept of conservation of volume. You cannot make something longer and keep the other dimensions the same. You actually can see the result of this effect with large wall tapestries which are only hung along the top (see Fig. 2.14). The material near the top is thus carrying all of the weight of the tapestry and is strained downwards (tensile) the most. In order to conserve volume, the tapestry must therefore shrink parallel to the top edge. As a result, the sides pull in slightly and there is wrinkling at the bottom (red arrows in Fig. 2.14). Some readers may have seen this wrinkling effect when trying to hang a bedsheet as a temporary window curtain. Although the top edge of the sheet may be straight, the bottom always wrinkles.

What happens is that beyond σ_{UTS}, microstructural processes in the material result in the tensile specimen quickly becoming much thinner. This is a process known as necking (see Fig. 2.15). The force required to continue deforming the specimen thus decreases. This means that the stress calculated using the original load carrying area, A_0, $\sigma = F/A_0$, decreases as we see in the stress–strain curves. If we were to calculate the stress based on the momentary force and cross-section area based on the Poisson effect, the so-called true stress, we would find that the stress increases all the way to final failure. Since most engineering applications only require knowledge of the stress–strain curve up to σ_{UTS}, or only the material strength, using the original area, A_0, provides enough accuracy. The stress–strain curves which we have been showing and are commonly used are aptly known as engineering stress–strain curves. For fundamental research on materials properties, the true stress and thus true stress–strain diagrams are used, which require more experimental effort to measure than described in Section 4.2. In this book, we will always refer to stress as the **engineering stress**.

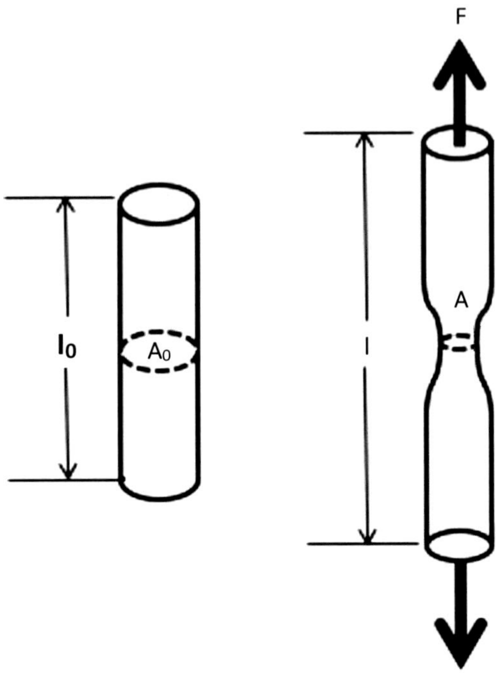

Figure 2.15 Necking of a tensile specimen. (a) Before loading. (b) Necking just before final failure.

Typical ranges of values for the yield strength, ultimate tensile strength and/or *E*-modulus of a number of classes of materials which one typically comes across in art and objects conservation are shown in Table 2.3. These values are far from all-inclusive and are only meant to provide a feel for mechanical properties data. One should refer to independent materials handbooks (see, for example, references [1–5]), as well as materials experts and eventually the supplier for mechanical properties data for a specific application.

As we discussed at the end of Section 2.2, many metallic engineering components and consumer products are not designed with the material strength in mind, but in terms of the amount of deformation. In principle, most components and products are designed so that they do not permanently deform when used properly, that is, their yield stress is high enough. However, if there is an accident, the product is designed to absorb energy by permanently deforming rather than breaking. An extreme example is a head-on automobile accident. One certainly would prefer that the bumper, and frame and engine bay permanently deform, rather than instantly breaking.

Table 2.3 Typical ranges of mechanical properties at room temperature for the several classes of materials

Material class	*E*-Modulus, *E* (GPa*)	Yield stress, σ_y (MPa)	Strength, σ_{UTS} (MPa)
Ceramics and glasses	50–400	—	50–1500 (in bending)
Metals and metal alloys (typical)	50–400	50–1000 (tension)	100–1200 (tension)
Aluminum	70	100–500	100–600
Steel	200	200–1000	350–1200
Polymers and plastics	<10	20–170	30–200 (tension; bending)
Wood	5–15 (bending)	—	50–120 (tension, parallel to grain)
Adhesives (shear properties)	<2	—	< 10 (shear)

*GPa = 1000 MPa.

Practical exercise 2.3

In this practical exercise, we will show how the concepts of stress, strain, and *E*-modulus come together in the selection of materials for a specific application.

A cylindrical supporting rod in a metal work of art needs to be replaced (see Fig. 2.16). The rod is loaded in tension. The rod will be painted, so the use of an original material is ethically not as important as the structural integrity of the object. The rod is 700 mm long and has a diameter of 12.7 mm.

The rod may not plastically deform, and it may not elongate more than 1.3 mm under an applied load of 29,000 N. The materials supplier has offered the four metals shown in Table 2.4 with the mechanical properties E-modulus, yield stress and strength. Which material would be selected and why?

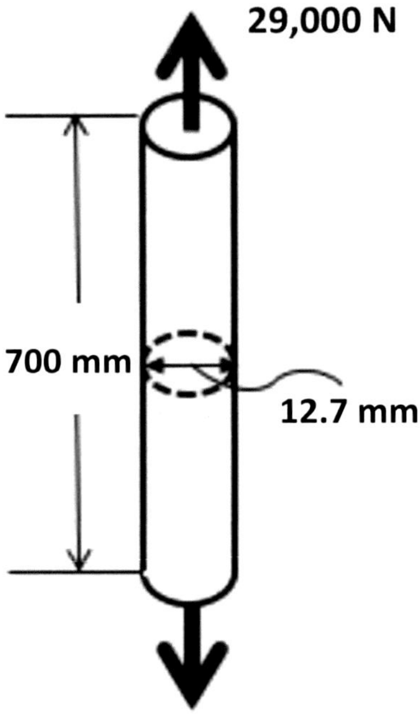

Figure 2.16 Cylindrical metal rod loaded in tension.

Table 2.4 Mechanical properties of candidate materials

Material	E-Modulus, E (GPa*)	Yield stress, σ_y (MPa)	Ultimate tensile strength, σ_{UTS} (MPa)
Aluminum alloy	69	255	421
Brass	120	345	421
Copper alloy	110	207	276
Steel	207	448	552

*GPa = 1000 MPa.

Solution

If one reads the problem carefully, it can be seen that the selection criteria for the material are based on strain. The material may not show any permanent strain (plastic deformation), and there is a limit on how much the rod may elongate under load, that is, there is a limit on the allowable strain.

The fact that the rod may not plastically deform means that the stress in the rod may not be higher than the yield stress, σ_y, of the material. We can calculate the stress in the rod using Eq. 2.3, noting that the load is already given to us in the units of force, that is, Newtons. The load-carrying cross-section of the rod is a circle, whose area $A = \pi r^2$, or in words, the area of a circle is the number, π, approximately 3.14, times the square of the radius (half of the diameter) of the circle.

$$A = 3.14 \times \left(\frac{12.7}{2}\right)^2 = 126.6\ \text{mm}^2, \quad \text{and thus}$$

$$\sigma = \frac{F}{A} = \frac{29{,}000\ \text{N}}{126.6\ \text{mm}^2} = 229\ \text{MPa}.$$

Looking at Table 2.4, one immediately sees that copper is not a suitable candidate because its yield stress, 207 MPa, is less than the stress in the rod. The other three materials will not plastically deform because their yield stresses are greater than that in the rod.

We now go to the second strain criterion. We see that the rod is 700 mm long and the allowable elongation is 1.3 mm. Using Eq. 2.6 or 2.7, we can calculate the allowable strain as

$$\varepsilon = \frac{(l - l_0)}{l_0} = \frac{\Delta l}{l_0} = 1.3\ \text{mm}/700\ \text{mm} = 0.0019.$$

In order to calculate the strain in a rod of each of the remaining materials, we need to calculate the strain in a rod made of each material under the stress of 229 MPa. In order to this, we need to rewrite Eq. 2.10 as

$$\varepsilon = \frac{\sigma}{E}$$

In this case, σ and E are known. We just calculated σ = 229 MPa, and we can find the E-modulus of each material in the supplier's Table 2.4. Note that the E-modulus values are most often given in GPa (giga-Pascal), which is 1000 times MPa. The strain in a rod of each material is then

$$\varepsilon(\text{aluminum}) = \frac{229 \text{ MPa}}{69{,}000 \text{ MPa}} = 0.0033$$

$$\varepsilon(\text{brass}) = \frac{229 \text{ MPa}}{125{,}000 \text{ MPa}} = 0.0019$$

$$\varepsilon(\text{steel}) = \frac{229 \text{ MPa}}{207{,}000 \text{ MPa}} = 0.0011.$$

Brass and steel fulfill the requirement that the strain cannot be more than 0.0019. One can thus select between these two materials. However, brass is right at the strain limit, so to be safe, one would probably want to choose steel for the rod.

We close this section by noting that the technical definition of the words, "elastic," "elasticity," "plastic" and "plasticity" have little in common with the definition of the word as used by artists, art connoisseurs and also the layperson for works of art. At the most, describing a work of art as elastic implies a sort of stretchiness, either physically or figuratively, while describing it as plastic implies something about the real or figurative form. Technically, elastic deformation only describes deformation which can be recovered when a load/ stress is removed from an object, while plastic deformation is permanent.

2.5 Ductile vs. Brittle Materials

In the previous two sections, we discussed the relationship between stress and strain (load and deformation), and introduced two terms for deformation, elastic strain which is recoverable, and plastic strain which is permanent. These discussions form the basis for the last two basic concepts and terms that we will define and discuss, the concepts of **ductility** and **brittleness**, and the difference between ductile and brittle materials.

Referring back to the schematic diagrams in Figs. 2.9 and 2.10, we see that there are many types of stress–strain curves depending on the material. Ceramics (Fig. 2.9) typically show only linear elastic behavior and their failure strains are extremely small, often less than 0.1%. The stress–strain curves for most ceramics are very steep and geometrically, there is very little surface area under the curves. Such materials with low fracture strains, little or no plastic deformation and very small areas under their stress–strain curves are known as **brittle** materials.

On the other hand, most metals show plastic deformation and will fail at strains from 25% to over 50%. Wood and textiles, Fig. 2.10, also often show permanent deformation, and in all three cases, there is a significant surface area under their stress–strain curves. Such materials with larger fracture strains, which show large amounts of plastic strain, that is, permanent deformation, and have significant surface areas under their stress–strain curves are known as **ductile** materials. Based on this large surface area, they are also referred to as **tough** materials. Many plastics, especially elastomers (Fig. 2.9) appear to have little area under their stress–strain curves if one thinks of the curves as ceramic stress–strain curves turned sideways. However, the failure strains for these materials can be several hundred percent. Thus, although the tensile strength of these materials might be low as compared with metals, the area under the stress–strain curves for elastomers is still much larger than for ceramics. Combined with the huge fracture strains, such plastics can certainly also be considered ductile.

There are, of course, many exceptions, and these will often be found among the many thousands of types of organic materials and plastics. Certain metals are actually considered to be brittle, while at extremely high temperatures, some ceramics as well as glasses can become very ductile. The important thing to remember is that there is a technical definition for ductility and brittleness.

2.6 Stress Concentrations

Up to now, we have discussed stress and strains which are based on the simple geometry where the applied force is evenly distributed over the load-carrying area (see Fig. 2.1). This will be useful for many applications in art conservation. However, there will be many situations in which this will not be the case. Of particular concern are objects with relatively sudden changes in geometry, such as holes, notches, sharply curved parts, and in the extreme, cracks (see Fig. 2.17). These sudden changes in geometry are known as **stress concentrations** or stress raisers. The stress at these locations is significantly higher than we would expect using Eq. 2.4.

The local stress at such stress concentrations can be calculated as

$$\sigma_{\mathrm{max}} = k_{\mathrm{c}} \times \sigma_{\mathrm{n}} \tag{2.11}$$

where σ_{max} is the maximum stress at the stress concentration, k_{c} the so-called **stress concentration factor**, and σ_{n} the so-called nominal stress, the stress, σ, we would have calculated with Eq. 2.4 using the cross-section load-carrying area without the notch.

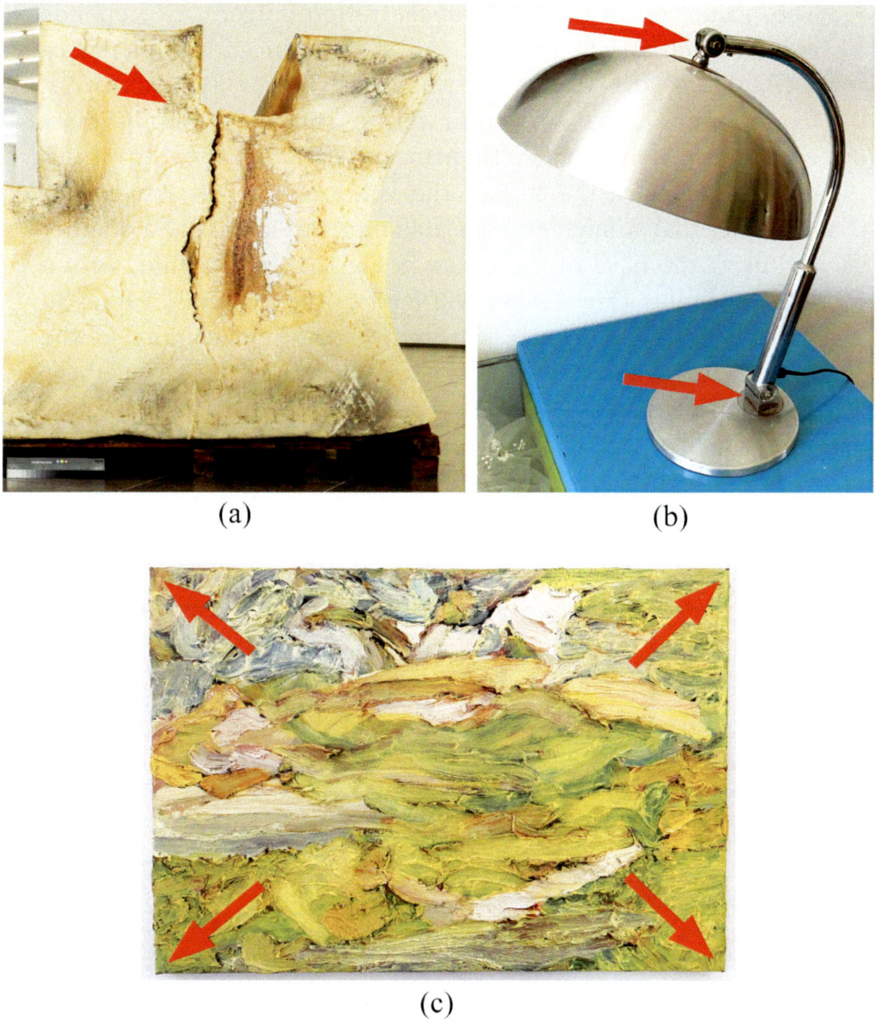

Figure 2.17 Stress concentrations (red arrows) in various objects. (a) Sharp corner in aging plastic foam material (part of "Unschlitt" (1977) by Josef Beuys, in the Hamburger Bahnhof Museum für Gegenwart in Berlin, Germany). (b) Lamp connections where most movement is expected. (c) Corners where painting canvasses are keyed ("Untitled (yellow))" (2003), Marie José Robben, 50 × 70 cm, www.robben.amsterdam). Note that on a detail level, sharp changes in the height of impasto can also work as stress concentrations.

The value of the stress concentration factor, k_c, depends on the geometrical form of the stress concentration. Figure 2.18 shows the range of values of k_c for typical forms. Graphs and tables of values for k_c can be found in most engineering handbooks. In principle, the "sharper" a stress concentration is, the higher the value of k_c. This means, for example, that a smaller hole diameter in Fig. 2.18a, a smaller radius at the base of the notch in Fig. 2.18b, or a fillet with a smaller radius (Fig. 2.18c) will lead to the higher k_c values in the ranges given. There is a limit to the applicability of the concept of stress concentrations, and that is the case where a crack is present in an object.

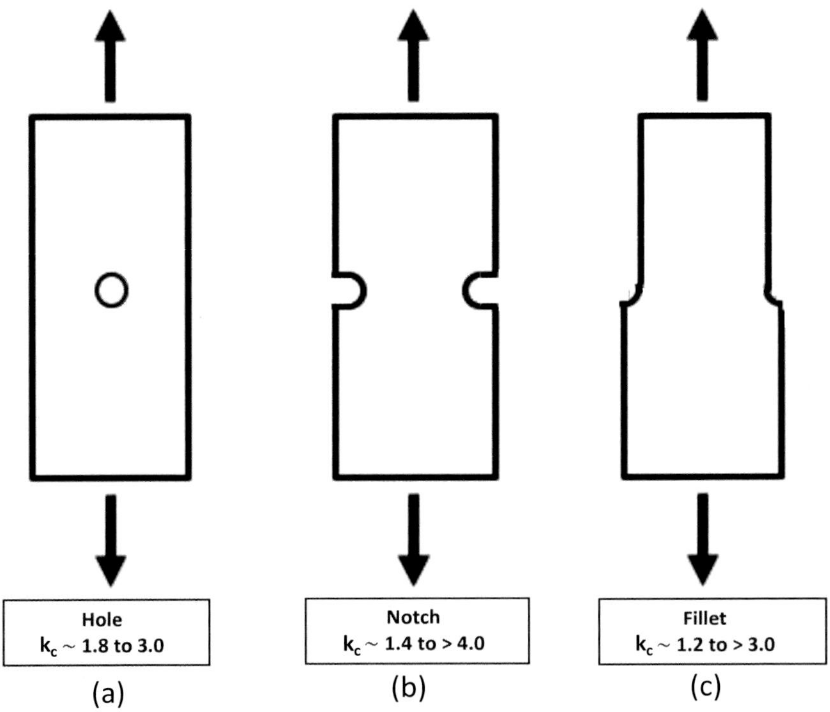

Figure 2.18 Range of values for the stress concentration factor, k_c, for typical forms of stress concentrations (stress raisers). The values of k_c depend on the dimensions of the stress concentration as related to the overall dimensions of the object.

A crack can be seen as a notch with an almost infinitely small radius. This would mean that the stress concentration factor, k_c, would be infinitely high. For reasons related to plasticity, this is not the case. The prediction of the failure of objects for such cases is the realm of the field of fracture mechanics. This will be discussed in Section 5.2.

2.7 Internal Stresses

The basic concepts of mechanical properties we have been discussing have been based on loads applied to an object by an external physical source. However, damage to objects can also be caused by so-called **internal stresses**. These are induced in objects by among others, changes in environment, in particular, large, "rapid" changes in temperature and humidity, or during the creation of the object or its restoration/repair through the improper use of heating devices. Those readers from the author's generation may have made his mistake of trying to rinse an old-fashioned silicate glass baking dish almost hot out of the oven under a stream of cold water and have it "explode" in his hands.

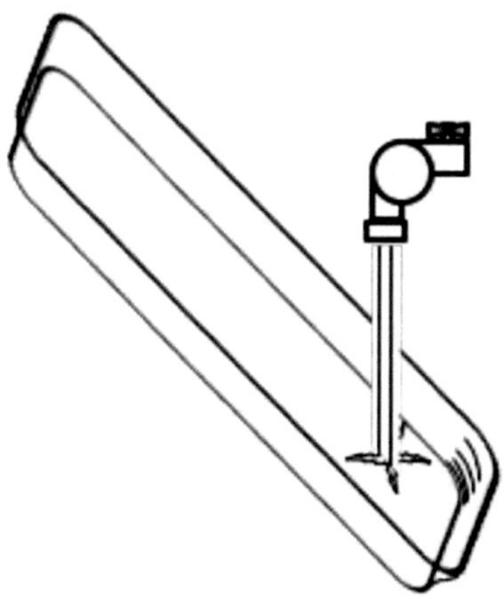

Figure 2.19 Thermal shock where cold water hits a hot glass baking dish.

The key to the development of internal stresses is that the object or parts of the object are fixed in such a way that the object as a whole cannot respond to the change. When the hot glass dish is suddenly placed under a cold stream of water, it undergoes what is known as thermal shock, a sudden and large local change in temperature where the cold water meets the hot glass (see Fig. 2.19). In technical terms, this is known as a high temperature gradient. Under the cold water, the material at the contact point wants to contract immediately. However, the rest of the glass is still hot and acts as a rigid frame. Because the material in the contact zone cannot contract, internal stress is quickly built up at the boundary between the hot frame and the local cold zone.

The amount of stress which is built up in the glass can be estimated using the concept of thermal expansion and Hooke's law (Eq. 2.10). When the temperature changes, almost all materials will experience a change in dimensions based on their coefficient of thermal expansion, written using the Greek letter α_T. Note that the units of α_T are mm per mm per degrees Celsius (mm/mm-°C) or mm per mm per Kelvin (mm/mm-K), that is, 1/°C (one over degree Celsius) or 1/K (one over Kelvin). Recalling the discussion of deformation in Section 2.2, this means that α_T is actually given in units of strain per unit temperature. The strain caused by a change in temperature is calculated by multiplying the thermal expansion coefficient, α_T, by the change in temperature,

$$\varepsilon = \alpha_T \times (T_2 - T_1) \text{ or} \tag{2.12}$$

$$\varepsilon = \alpha_T \times \Delta T \tag{2.13}$$

where ε is strain, α_T the coefficient of thermal expansion for the material, T_1 the starting temperature, T_2 is the final temperature, and Δ the change.

For normal glass, α_T is approximately 0.000009/°C or in scientific notation, 9×10^{-6}/°C. If the temperature of the glass was 150°C and that of the cold water was 20°C, then the contact point would be trying to shrink with a strain of

$$\varepsilon = 0.000009/°C \times (20°C - 150°C) = -0.00117.$$

Because the glass at the contact point cannot, in fact, shrink, an internal stress is built up. We estimate this internal stress using Hooke's law (Eq. 2.10) assuming that ceramics only behave elastically. We have

$$\sigma = E \times \varepsilon \qquad \text{(Eq. 2.10)}$$

where ε = –0.00117, which we just calculated. The *E*-modulus for glass is approximately 70 GPa or 70,000 MPa. Substituting these two values into Eq. 2.9 gives us a stress of σ = 81.9 MPa.

If we compare this value of stress, 81.9 MPa, or the value of strain which the glass would have expanded for this temperature, 0.00117, with values of the mechanical properties of silicate glass in the literature, we will find that our values are pretty much equal to those for the failure of glass. In fact, the glass baking dish or pretty much any standard glass cup or dish could also fail suddenly at lower "shock" temperatures. Because ceramics and glasses are so brittle, they are notoriously difficult to test, and the scatter in values for strength and failure strain values is huge. This is because such materials are very susceptible to microdefects, essentially very small stress concentrations or cracks (see the previous Section 2.6). It is therefore not uncommon that a cheap tea glass will also crack under a stream of cold water, even if the water in the glass was under 100°C.

A better-known example of the effect of internal stresses is the crack in the Liberty Bell in Philadelphia, Pennsylvania, in the USA. The crack is actually a typical example of many such cracks in cast metal bells around the world. If a bell, or any metal casting for that matter, is not properly and evenly cooled, the parts that cool faster will, as in our example of the glass baking dish, develop internal tensile stresses. In the materials engineering world, these are also known as **residual stresses**. Recall that metals are much more ductile than ceramics and glass. Thus, the strains which are being resisted by the surrounding warmer parts will not lead to immediate fracture. The finished bell is, however, pre-loaded with the residual stress. Any external stress applied to the bell, such as ringing it, will be added onto the residual stress. What occurs in most cases, is that the bell begins to show hairline cracks. These grow with time until one large crack forms. We will discuss this cumulative process of damage known as fatigue in Section 3.2.

Similar production problems can result in high internal stresses in plastics. This can lead to the formation of a whitish discoloration within the materials. A closer look shows that this is a network of cracks known as crazing, which are caused by the realignment and debonding of polymer chains due to the internal stresses. This network of cracks can eventually lead to cracking of the object, but initially, the network does not reach the surface and cannot be felt with the bare finger as a crack could be.

Another familiar example of cracking due to internal stresses is the formation of craquelure on paintings, in particular, traditional oil paintings. Most paints want to shrink as they dry. For certain types of paints, this can result in so-called alligator cracking, where the paint layer does not bond well with the surface underneath. It then pulls apart locally as it dries. You see this behavior also when rain or wash water dries on an automobile that has just been waxed.

The most common form of craquelure is, however, the distinct, sharp crack patterns which are found on many classical oil paintings, see Fig. 1.1b in Chapter 1, or sharp cracks which on found on more modern paintings painted with thick impasto. This type of cracking occurs as a more complex process through time. Initially, the top part of a paint layer dries first, before the underlying part. The top "skin" thus develops a residual tensile stress. Fresh (oil) paints are fairly ductile, so that cracking does not immediately occur. Over the years, the paint ages and becomes more brittle. At some point in time, the paint layer will begin to crack due to a combination of the residual stresses and any external changes in temperature and/or humidity, and the increased brittleness of the paint. This is the main result of the extensive research conducted by Marion Mecklenburg on the mechanical properties of paints (see, for example, references [6–8]). This is often accompanied by "cupping" of the paint flakes. The top surface can be pulled together due to the higher residual stresses there. This is the same effect that one sees in a dried-out riverbed. The phenomenon of cupping will be discussed further in Section 5.2 with reference to Fig. 5.8. Note that there is a misconception that craquelure, once established, can continue to grow. This will also be discussed in Section 5.2.

References

1. Gauthier, M. M., editor. (1995). *Engineered Materials Handbook Desk Edition* (ASM International, Materials Park, OH, USA).
2. Davis, J. R., editor. (1998). *Metals Handbook Desk Edition*, 2nd ed. (ASM International, Materials Park, OH, USA).
3. Kuhn, H., and Medlin, D. editors. (2000). *ASM Handbook Volume 8: Mechanical Testing and Evaluation* (ASM International, Materials Park, OH, USA).
4. Osswald, T. A., Baur, E., Brinkmann, S., Oberbach, K., and Schmachtenberg, E. (2006). *International Plastics Handbook 4E: The Resource for Plastics Engineers* (Carl Hanser Verlag, Munich).
5. U.S. Department of Agriculture. (2007). *The Encyclopedia of Wood* (Skyhorse Publishing, Inc., NY, USA).
6. Mecklenburg, M., Tumosa, C., and Erhardt, D. (2004). The changing mechanical properties of aging oil paints. In: *Materials Research Society Symposium: Materials Issues in Art and Archaeology 7*, eds. Vandiver, P. B., Mass, J. L., and Murray, A., (Materials Research Society, Warrendale, PA, USA) pp. 13–24.
7. Mecklenburg, M. F., *Determining the Acceptable Ranges of Relative Humidity and Temperature in Museums and Galleries, Part 1, Structural Response to Relative Humidity* (Smithsonian Museum Conservation Institute, Washington, D.C., USA), https://repository.si.edu/handle/10088/7056 (accessed January 16, 2025).

8. Mecklenburg, M. F., *Determining the Acceptable Ranges of Relative Humidity and Temperature in Museums and Galleries, Part 2, Structural Response to Temperature* (Smithsonian Museum Conservation Institute, Washington, D.C., USA), https://repository.si.edu/handle/10088/7055 (accessed January 16, 2025).

Chapter 3

Dynamic Mechanical Properties

In the previous chapter, we discussed the terminology associated with so-called static loads/stresses. These are loads on an object which do not change in time. Of course, the load has to be initially applied, but they are applied slowly, relatively speaking, such as sitting down in a chair, hanging a chandelier, or placing a stone sculpture on the floor in a museum or on the ground in a public square. Once the loads are applied, they remain constant.

In this chapter, we will look at the basic terminology for dynamic loads, that is, loads which do change in time. These include loads which can change very quickly, commonly known as impact or shock loads, or loads which vary cyclically with time, commonly known as cyclic loads.

3.1 Shock

Dropping an object, hitting a nail with a hammer, a truck hitting a speed bump too fast or in the extreme, a construction company blasting a new tunnel, are examples of loads which are applied in an extremely short period of time, milliseconds, as shown schematically in Fig. 3.1a. Such loads are known as

Art Conservation: Mechanical Properties and Testing of Materials
W. (Bill) Wei

ISBN 978-981-5129-63-2 (Hardcover), 978-1-003-61644-3 (eBook)
www.jennystanford.com

shock or impact loads. Because the loading is so fast, the material deformation will not just follow the stress–strain diagrams up and down once as we saw in Chapter 2. The material will behave like a spring that has been pulled and then suddenly let go. The material will deform back and forth with several peaks as shown schematically in Fig. 3.1b. This can be thought of in terms of an earthquake followed by a number of aftershocks. Unlike an earthquake, however, the cyclic deformation due to a shock load dies out very quickly.

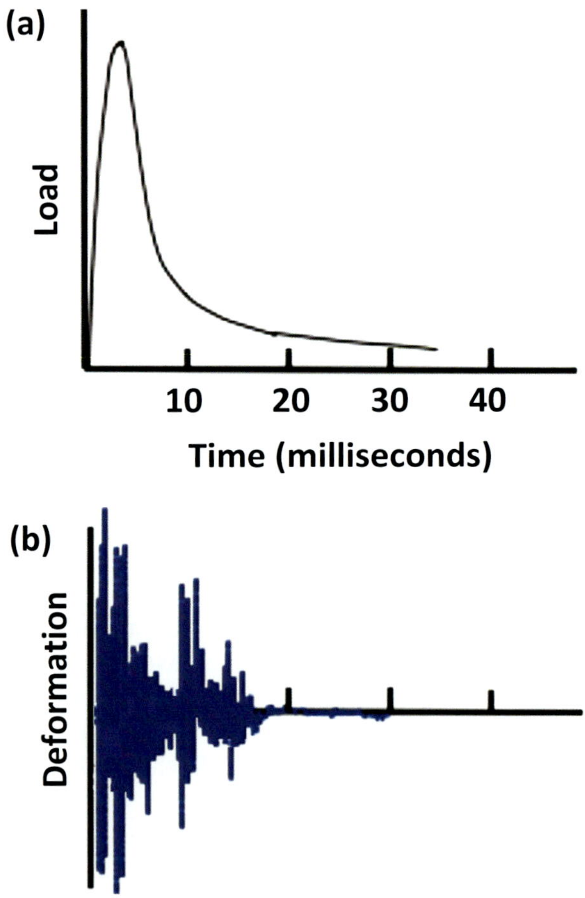

Figure 3.1 Shock loading. (a) Schematic plot of the load versus time during an impact load. (b) Schematic material response to impact load.

For the purposes of this book, it is sufficient to say that the object will suffer damage or fail according to the stress–strain diagram for the materials it is made of. If the local stress is higher than the yield stress, then the object will be permanently deformed, and if the local stress is higher than the materials strength, then it will severely crack or break.

Having said that, it must be noted that the prediction of the deformation and failure behavior of an object for a given shock/impact loading is much more difficult and complex than for static loading. Materials which are deformed at very high loading rates do not react in the same way as materials which are statically loaded. A shock load puts a large amount of energy into a material in such an extremely short time, that the material does not have enough time to react and deform the way it would under static loading. For example, many otherwise ductile materials will fail in a brittle manner under shock loading. The damage which is caused will also depend on the local stresses at the impact point, the geometry of the object, and the orientation at which it falls. As you may already know from experience, dropping a smartphone from a given height and having it land flat produces a different work of art on the screen than having it land on its corner, even though the drop height may have been the same in both cases.

Because of this complexity, it is difficult to describe a shock load in the terms of stress and strain as these terms are used for static loads. A number of parameters which are used to describe shock loads as well as vibration loads will be discussed in Section 3.6. For transport applications, the use of "*g*" and "*g*-force" to describe shock is probably the most familiar. As will be discussed, relating such parameters to actual damage and determining "safe" limits is extremely difficult, and still a major subject of ongoing research in the engineering industry.

We conclude this section and introduce the following section by referring back to the cyclic deformation response to a shock load which was shown schematically in Fig. 3.1b. The term used by most vibration and acoustics engineers and consultants to describe this cyclic deformation behavior is "vibration." While this is true, the use of the term "vibration" in this way often leads to confusion and misunderstanding in the art conservation field, where the terms shock and vibrations are very often used synonymously. As we just discussed, the vibrations which occur after a shock load have a duration of only a few milliseconds. On the other hand, **vibrations** which are most often of concern to art conservators can last for hours, such as those during art transport, or a rock concert in or near a museum, or up to weeks at a time for vibrations caused by construction work or heavy traffic near a museum. We will discuss the term, vibrations, in more detail in Section 3.3. It is, however, important for professionals in the conservation world to remember that *shock and vibration are two completely different kinds of loadings* with different consequences for an object.

3.2 Cyclic Loading: Definitions

In Chapter 2 and Section 3.1, we discussed two types of loading which occur once in time, static and shock/impact loading. In this section, we introduce the concept of **cyclic loading**. Examples of cyclic loads are

- the seasonal thermal and humidity changes which works of art have been exposed to throughout a history without climate control (Fig. 3.2a),
- the operation of a passenger airplane over periods of years, where it takes off and lands several times a day (Fig. 3.2b),
- a pump in an oil field in Texas which is constantly moving up and down (Fig. 3.2c),
- the load on the wings of a passenger airplane as it flies through turbulence (Fig. 3.2b),
- the vibrations in a truck as it drives over a rough road (Fig. 3.2d), and
- (loud) music (Fig. 3.3e).

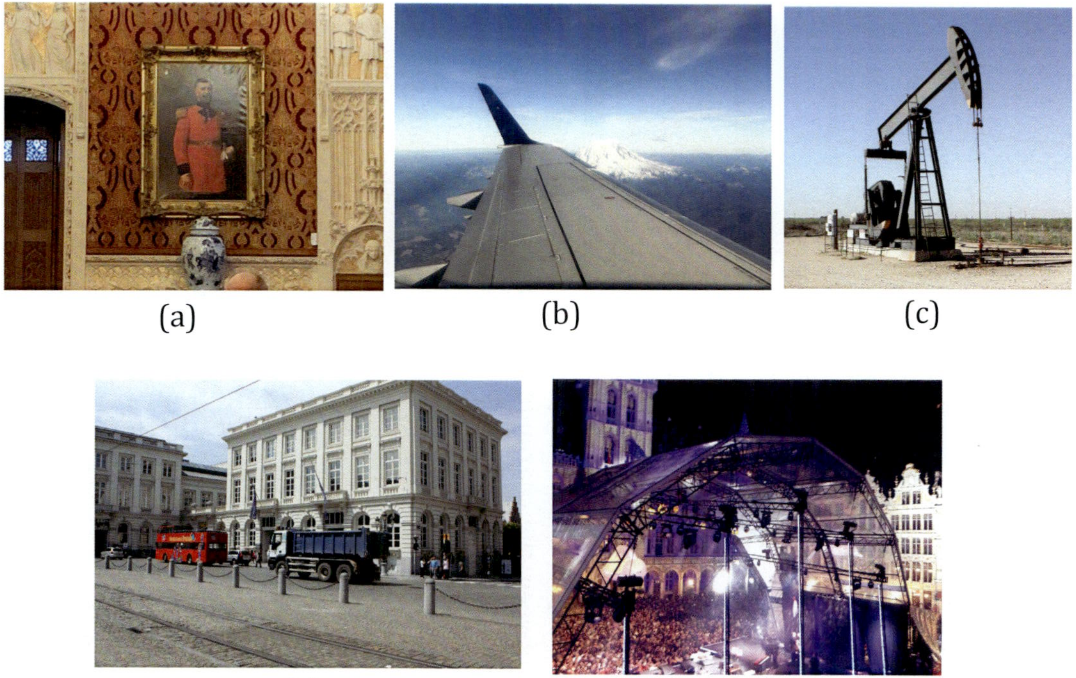

(a) (b) (c)

(d) (e)

Figure 3.2 Examples of cyclic loading in daily life. (a) Seasonal temperature and humidity changes in uncontrolled climates. (b) Cyclic loading of aircraft per flight and through turbulence. (c) Oil pump in Texas. (d) Truck on a cobblestone road in front of two museums in Brussels. (e) Rock concert on the historic Grote Markt in Brussels.

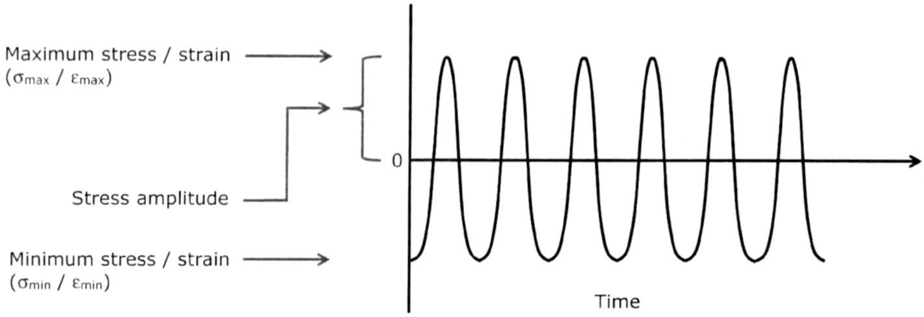

Figure 3.3 Description of a simple cyclic load in the form of a sine wave.

The simplest form of a cyclic load is shown schematically in Fig. 3.3. During cyclic loading, the load/stress changes regularly in time, cycling back and forth between a minimum and a maximum value. (One can also have situations where the strain is cycled between a maximum value and a minimum value.) The **amplitude** of the cyclic load is defined as half the distance between the maximum and minimum. In Fig. 3.3, that happens to be the distance between zero load and the maximum. The height of the wave from the maximum to the minimum (the so-called peak-to-peak value) is thus twice the amplitude. The number of times the load cycles in a given amount of time is known as the **frequency**. For the examples given in Fig. 3.2, that could be one to several times a year for seasonal temperature and humidity cycles for the painting without climate control (a), several cycles (flights) per day for the airplane flights, or several cycles per second for turbulence (b), several cycles per minute for a Texan oil pump in operation (c), to many cycles per second for the truck (d) and the music (e).

Figure 3.3 was drawn with the cyclic load being symmetric around zero stress or strain (see also Fig. 3.4a). This is known as full-stress reversal where the stress cycles between the same tensile and compressive value. However, in many cases, the loads cycle around a mean which is not zero. The load cycle can be completely tensile (Fig. 3.4b), completely compressive (Fig. 3.4c), or a mix of both (Figs. 3.4ad). This can have some influence on the fatigue life of an object. Engineers describe these variations using a so-called ***r*-ratio**, the minimum stress level divided by the maximum stress level, or in equation form,

$$r = \frac{\sigma_{\min}}{\sigma_{\max}} \tag{3.1}$$

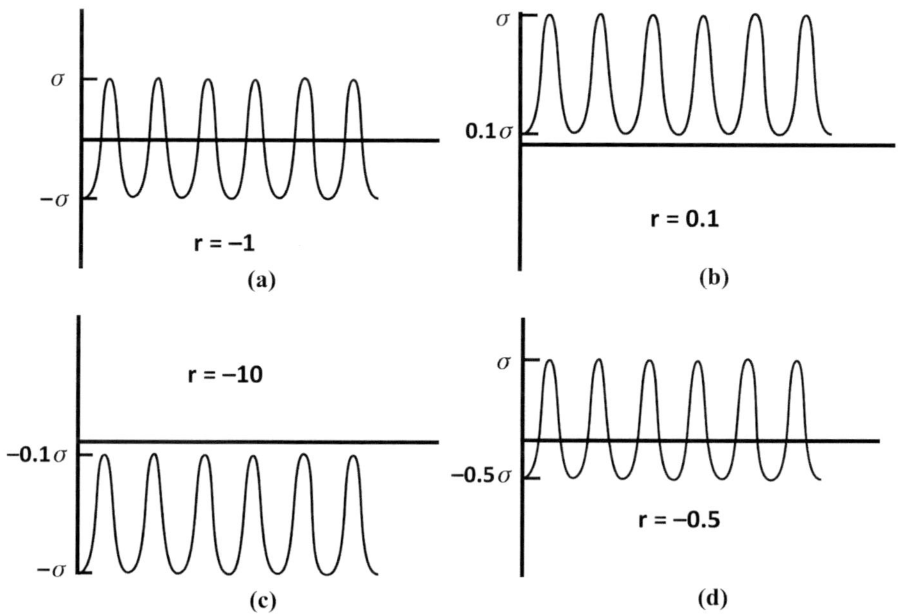

Figure 3.4 Cyclic loads around different mean stresses. (a) Tension-compression (full reverse) loading around zero mean stress. (b) Tension-tension loading. (c) Compression-compression loading. (d) Tension-compression loading, generalized version of (a).

The r values for the four combinations of minimum and maximum stresses are shown in Figs. 3.4a–d. An example of a full cyclic load reversal from tension to compression (Fig. 3.4a) is the surface of a painting when it vibrates (see Fig. 3.5a). On the other hand, the beams of a balcony in a historic house (Fig. 3.5b) have to carry their own weight, so that the underside of the beams will always be under a slight tensile load (see (1) marked in Fig. 3.5b). However, as people walk across the balcony, the underside of the beams will be loaded in tension-tension (Fig. 3.4b) first to a maximum based on the weight of the people, and then return to a minimum based on the weight of the beam. The top surface of the beam will see the opposite loading, compression-compression (Fig. 3.4c; see (2) marked in Fig. 3.5b).

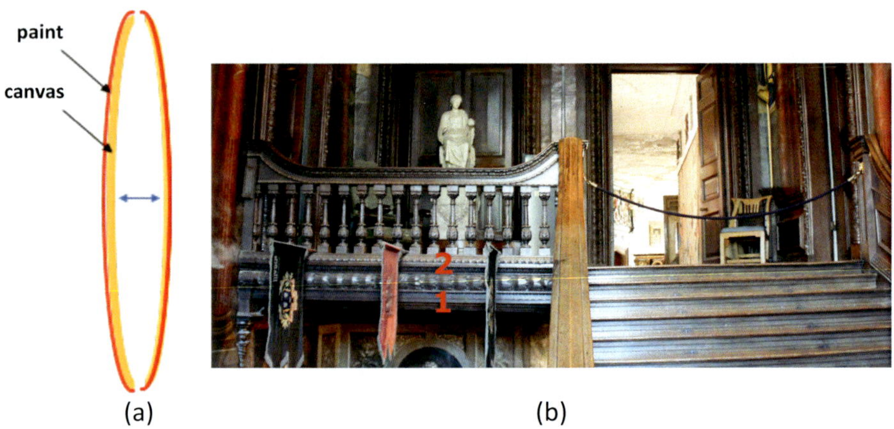

Figure 3.5 Examples of cyclic loads with different mean stress levels. (a) Vibration of a painting: the paint layer goes through full reverse loading. (b) Bending of a balcony support beam. The lower surface (1) undergoes tension-tension and the upper surface (2) compression-compression loading as visitors walk on and off of the beam.

3.3 Cyclic Loading: Damage Accumulation

One of the most important differences between static and shock loads on the one hand, and cyclic loads on the other, is that damage due to cyclic loads is cumulative. The initiation of damage may take a large number of cycles and will often be in the form of micro-cracking, invisible to the naked eye. The damage accumulates, whereby the micro-cracks grow and one main crack develops until the material or object fails. This process of cyclic failure is known in the engineering world as **fatigue**. The danger of fatigue is that it can occur at maximum cyclic stress levels which are well below the strength of the material of which the object is made.

Practical exercise 3.1

The reader can perform a simple experiment with a paper clip to demonstrate this cumulative damage effect. It would be impossible to pull the paperclip apart with one's bare hands/fingers in one pull or bend. However, if one continually bends it back and forth, the paper clip will eventually break after a number of bending cycles.

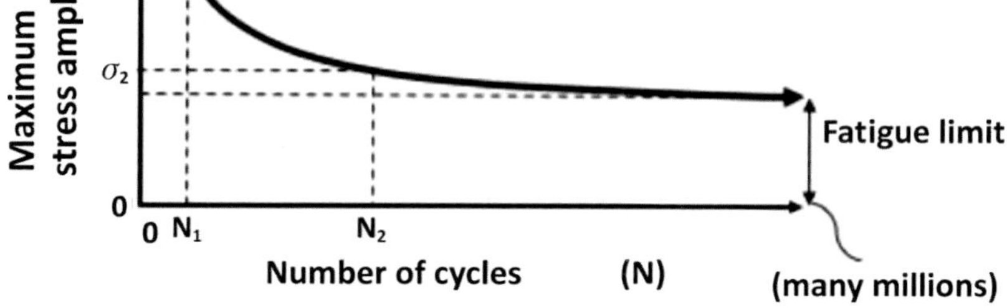

Figure 3.6 Prediction of the life of a material or object using a Wöhler (*S–N*) diagram. The vertical axis can also be given in units of maximum strain or strain amplitude.

One of the main contributions to understanding fatigue were studies conducted by the German railroad engineer, August Wöhler, in the second half of the nineteenth century. He was trying to understand why the axles of freight cars were breaking during long trips, even though they were not overloaded according to static design criteria. He found that the cyclic loads of daily travel, such as a train travelling over switches or the expansion gaps between rails led to cracking and eventual "sudden" failure of the axles.

He and his successor developed what is now known as the Wöhler diagram or *S–N* diagram to experimentally predict the life of components during cyclic loading. A schematic Wöhler diagram is shown in Fig. 3.6. The vertical axis is the cyclic stress level, often given as the maximum stress during cyclic loading or the stress amplitude (recall Fig. 3.3). The horizontal axis gives the number of cycles, *N*. Materials or objects are tested at different cyclic stress levels, and the number of cycles is determined when they fail. The curve in Fig. 3.6 thus shows the life of the material or component in numbers of cycles as a function of the cyclic stress level.

This diagram can be read as follows. For a maximum stress on the material of σ_1, one draws a line horizontally to the curve (dashed line). Where it meets the curve, one reads straight down following the vertical dashed line until it meets the horizontal axis. There one finds the number of cycles that the material or object would function until it failed, N_1. Similarly, if the object were loaded with a maximum cyclic stress of σ_2, the life of the object in numbers of cycles would be N_2. You can see that at the lower maximum cyclic stress, σ_2, the object would function for a larger number of cycles, N_2 than at the higher cyclic stress, σ_1. We say that the fatigue life of the object is very short at high cyclic stress levels, and longer at lower cyclic stress levels.

The upper limit of the Wöhler diagram at the upper left of Fig. 3.6 is, of course, the material strength. The life of the material is essentially a quarter of a load cycle. Earthquakes are essentially short-term cyclic loads at this end of the diagram.

At the other end of the diagram, the curve flattens out at low stress levels and large numbers of cycles, in particular, for many metals. Below a certain cyclic stress level, the material essentially will not fail. This cyclic stress level is known as the fatigue limit, or fatigue strength.

For basic materials testing as described above, failure is defined as the fracture of the test specimen. For many materials, including most commonly used metals, N can go into the millions of cycles. However, there are many consumer and industrial products for which designers and engineers use other definitions of failure. Those millions of cycles are then incorporated into design criteria which determine, for example, how many flights a passenger airplane may take until it must be inspected, or how long the guarantee for a consumer product should be. These are generally based on when the product no longer can be safely used as it was designed or advertised to be used.

In art conservation, the definition of failure is based on more subjective concepts of aesthetics, conservation ethics and economics. Loss of paint from a canvas during a museum loan, or further crack growth in a wooden object due to climate changes could be definitions of failure for works of art. N in these two cases could then be defined as the number of times an object is allowed to go on travel, or the number of climate cycles an object may be exposed to before damage occurs. The fatigue limit for a work of art could be defined as the stress below which no visible damage occurs before the next restoration.

3.4 Cyclic Loading: Internal Stresses

In Section 2.7, we introduced the concept of internal stress in the context of static loading. Cyclic internal stresses also play an important role in the aging and degradation of objects. Such cyclic internal stresses are the result of cyclic changes in temperature and/or humidity. The amplitudes of the stress cycles can also be estimated using Eq. 2.12/2.13 in combination with Hooke's law, Eq. 2.10 for materials exhibiting elastic behavior.

A classic example of the effects of cyclic internal stresses is the cracking of wooden panel paintings that have been cradled. Cradling was originally performed to prevent the bending of panel paintings due to changes in the climate. This leads to internal stresses on the outer fibers of the paint layers and the wood panel because the wood cannot respond to changes in climate. The surface of the cradled panel painting will react first to any (cyclic) climate change while the interior is not yet "aware" that anything has changed. For example, a decrease in humidity will first reduce the moisture contact of the surface layers of the panel, causing those layers to want to shrink. However, they are prevented from shrinking by the cradled subsurface material that still contains moisture and does not yet react to the change in humidity. The surface thus comes under internal tensile stress. In the reverse direction, an increase in humidity will put the surface in compression as it tries to swell. This cyclic loading will eventually lead to the formation of surface cracking, and eventually to further crack growth (Crack growth will be discussed further in Section 5.2.3.).

Even without cracking, the phenomenon of cyclic humidity changes is familiar to readers who live in countries with major seasonal changes (winter/summer, monsoon/dry). Entry doors to homes or doors to cabinets tend to jam either in winter or summer due to some combination of swelling or shrinkage of the various components, which depends on the specific construction of the door/frame/cabinet and the climate conditions inside and outside the residence. Over the years, some of the swelling or shrinkage becomes permanent, which results in always having to force the door due to the resulting friction between the door and the door frame (see Section 5.4 for more on friction and wear). That need to continuously (and cyclically) force the door can also lead to permanent bending deformation of the door.

3.5 Cyclic Loading: Vibrations

Vibrations are one form of cyclic loading which are of particular concern for museum and collection managers. The principles which were discussed for cyclic loads in Sections 3.2 and 3.3 apply fully for vibrations. It is thus important to remember that if one is confronted with a vibration issue and wants to know the allowable level for the vibrations, one must also have an idea of how long they will last. A safe limit for vibrations thus consists of a level AND a duration.

A number of concepts specific to vibrations are now introduced in this section. In Section 3.2, we introduced the concept of frequency for cyclic loading, that is, the number of load cycles in a certain period of time. The frequencies of vibrations are often expressed in units of Hertz (Hz), where 1 Hz = 1 cycle per second. The second hand on a clock ticks at a frequency of 1 Hz. The term, Hertz, may be familiar as the description of electric current coming out of the plug in the wall, 60 Hz for the United States and 50 Hz for Western Europe.

The concepts of frequency and amplitude for cyclic loads are demonstrated graphically for the case of vibrations in Figs. 3.7ab. In both figures, the vertical axis is the level of the cyclic load/vibration, that is, the amplitude. The horizontal axis is time in seconds. In both figures, a cyclic load/vibration of low frequency of 5 cycles per second, 5 Hz, can be seen, and one of higher frequency, 39 cycles per second, 39 Hz. In Fig. 3.7a, the low frequency cyclic load has a higher amplitude than the higher frequency cyclic load. The opposite is true in Fig. 3.7b.

Practical exercise 3.2

This exercise demonstrates the concepts of amplitude and frequency for vibrations. However, the principles are equally applicable to cyclic loading in general. For the musicians

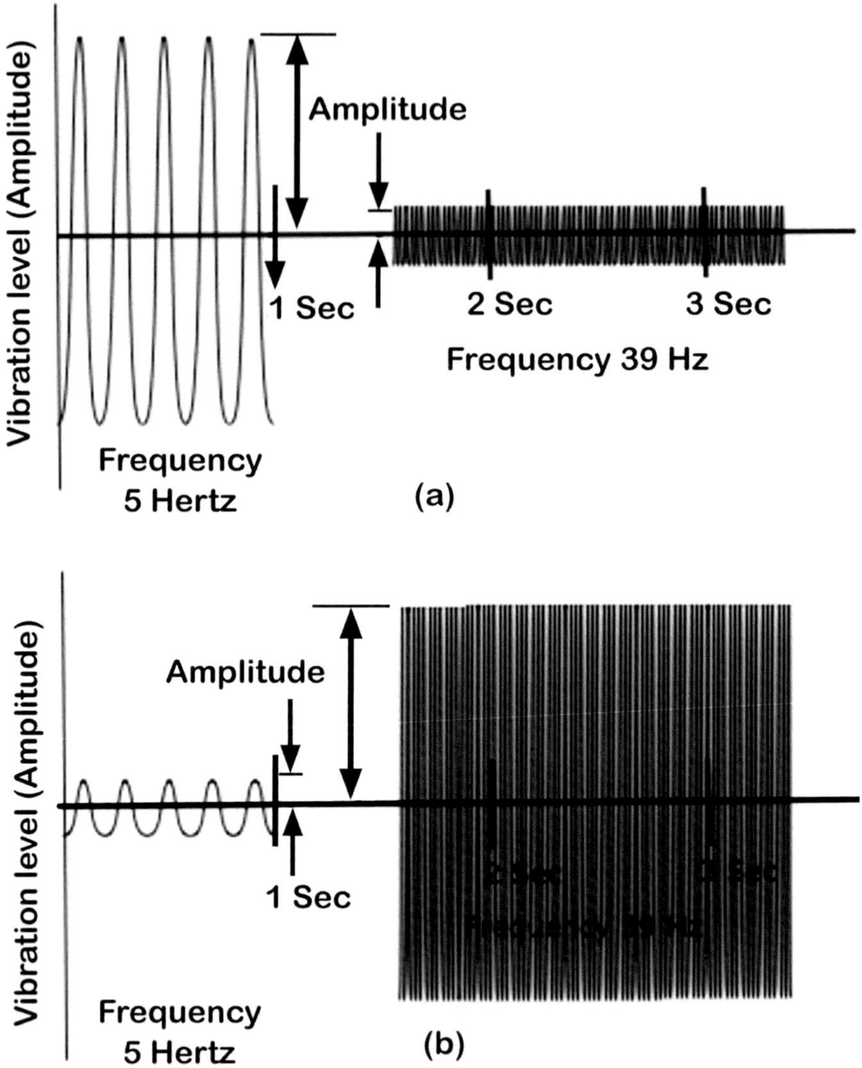

Figure 3.7 Visualization of vibration frequencies and amplitudes. (a) Low-frequency vibration of 5 Hz at high amplitude, and higher frequency vibration of 39 Hz at low amplitude. (b) Low frequency vibration of 5 Hz at low amplitude, and higher frequency vibration of 39 Hz at high amplitude.

among the readers, in particular, those who can sing, you can get a feel for amplitude and frequency by using the analogy of the low frequency vibration on the left of Fig. 3.7a or Fig. 3.7b being a bass in a choral group, and the high frequency vibration on the right side of Fig. 3.7a or Fig. 3.7b as a soprano. On the left side of Fig. 3.7a, the bass sings a very low note as hard as he can. That is a low frequency note sung with a high amplitude. On the right side of Fig. 3.7a, the soprano answers softly, that is at high frequency and a low amplitude. In Fig. 3.7b, the bass also responds softly, that is, with a low-frequency note at a low amplitude. In the finale on the right side of Fig. 3.7b, the soprano ends forcefully by singing a high frequency note at high amplitude.

There is no strict definition for when a cyclic load is a vibration or "just" a cyclic load, but as a rule of thumb, cyclic loads which have frequencies above 1-5 Hz into the thousands of Hz are classified as vibrations. As we noted in Section 3.1, an object also reacts to a shock load by vibrating, but that is for an extremely short period of time in terms of milliseconds. In this section, we have been using the terms cyclic loading and vibration primarily to discuss the loading and eventual damage of objects where the loading goes on for significant amounts of time as in our examples in Fig. 3.2. We therefore again remind the reader in the art conservation world, that shock and vibration are two completely different types of loading, and not to use these two terms synonymously.

3.6 Units for Shock and Cyclic Loading

In the engineering world, a number of parameters are used to describe shock and cyclic loads. These are briefly defined in Table 3.1 and described in the text below.

3.6.1 Stress σ and Strain ε

Fatigue data for a majority of industrial and consumer products are presented in terms of stress and strain. These two terms were described in Chapter 2. The exception is for vibrations, for which one of the other parameters in Table 3.1 is used in the literature.

3.6.2 *g* and *g*-Force

The reader who often deals with transport issues, may already be acquainted with the terms, "*g*" and "*g*-force," the second of which is a misnomer. As we saw in Section 2.1, "*g*" is actually the acceleration due to gravity on Earth with a value of 9.8 m/s^2. It is used quite often in the conservation literature, based on its use in transport packaging. It is also used as a sort of short hand to describe the acceleration caused by any shock load, no matter what the direction (gravity works downwards). If an object is shocked with 2.4 *g*, it experiences an acceleration of $2.4 \times 9.8 = 23.5$ m/s^2.

Table 3.1 Engineering terms and measurement parameters commonly used to describe shock/impact and cyclic loads

Parameter	Brief definition
Stress, σ	See Chapter 2; MPa or N/mm^2
Strain, ε	See Chapter 2; mm/mm (dimensionless)
g	Acceleration due to gravity with the value 9.8 m/s^2 (on Earth)
g-force	This term should not be used. It is a misnomer and technically incorrect. *g* is an acceleration, NOT a force
mm/s	Velocity
mm	Deformation/displacement
mm/mm	Strain
mm/s^2	Acceleration: Note that *g* is a value for a specific form (Earth's gravity) of acceleration
ASD	Acceleration spectral density
PSD	Power spectral density

3.6.3 Velocity (mm/s)

On the other hand, the construction industry and the building standards for most countries use velocity (mm/s) to describe the shock waves from blasting and other shock and vibration loads from heavy construction which arrive at nearby buildings. Note that this is not the velocity at which the wave moves from the shock/vibration source to the building, but how fast the ground moves up and down as the wave passes by (recall Fig. 3.3). If one throws a stone into a still pond, the ripples (waves) that are created move outward from the stone at some propagation velocity. However, the velocity we are talking about here is how fast a leaf or a bug on the water moves up and down as the wave passes by (see Fig. 3.8).

If one uses velocity to describe shock or vibrations, one must specify the frequency or frequency range. Most building codes related to vibration are based on such velocity–frequency relationships. This is due to the fact that in terms of energy, velocity is related to materials deformation. The kinetic energy of a vibration, $1/2\ mv^2$ is converted to deformation energy, the area under the stress–strain curve of a material, which was shown schematically for example in Fig. 2.8.

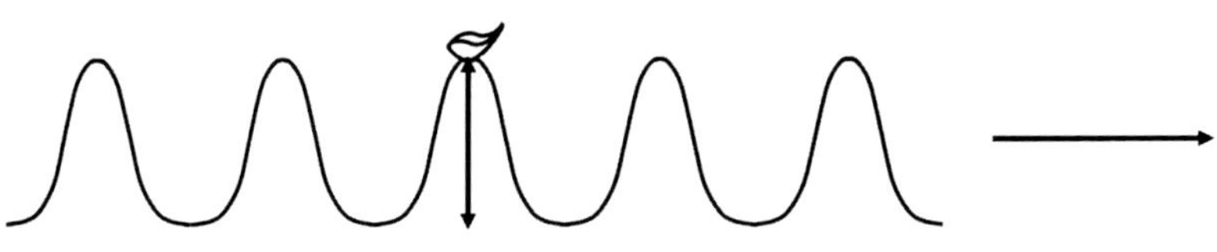

Figure 3.8 The leaf moves with changing velocity (Table 3.1) up and down as the wave moves by horizontally at some propagation velocity. The term velocity in the main text refers to the motion of the leaf.

3.6.4 Displacement (mm)

In order to describe deformation and damage during vibrations it would be useful to be able to use deformation/

displacement (mm). However, this is not often done since, for buildings, a large number of sensors (see Chapter 4 over testing) is required, and for objects of art, attaching a sensor is ethically not permitted.

Mathematically, the deformation can be calculated from the equation for the shock wave by so-called integration. Based on this, you can roughly estimate the deformation as the velocity (mm/s) divided by the frequency (1/s), or in equation form

$$d = \frac{v}{2\pi \times f} \tag{3.2}$$

where d is the maximum deformation in a vibration cycle, v is velocity in mm/s, $\pi \sim 3.14$, and f is the frequency per second.

3.6.5 Acceleration (mm/s²)

Acceleration (mm/s^2), often in the form of the parameter "g", is often used to monitor shock, as well as vibrations. For engineers, it is an easier unit to use because that is the unit which many sensors are designed for, which is why the sensors are called accelerometers. However, relating acceleration to deformation is extremely difficult. You can calculate deformation from acceleration by mathematically integrating twice, first to get velocity, and again to get deformation. Acceleration is thus a step further from deformation than going from velocity to deformation.

One can convert acceleration measurements to velocity using an equation similar to Eq. 3.2,

$$v = \frac{a}{2\pi \times f} \tag{3.3}$$

Note that the frequency must also be mentioned in both equations whether one uses acceleration or velocity.

3.6.6 Power Spectral Density (PSD or ASD)

A further parameter used for describing shock loads is the power spectral density (PSD). Shock loads, though very short, contain a large range of (vibration) frequencies. The PSD provides information about the distribution of energy among those frequencies, showing which frequencies are the strongest. This has relevance in terms of the resonance of objects which will be discussed in Section 3.7. Note that the word "Power" does not have the physical meaning you may be accustomed to when referring to an automobile or tool. It refers to the mathematical calculation used to obtain it, for example a number squared or cubed. For shock (and vibration) work, the acceleration is often used as the physical parameter in a PSD. The PSD is then referred to as an acceleration spectral density (ASD).

3.6.7 Which Parameter?

Because different engineering fields use different units to describe shock and vibration, it is not surprising that there is some confusion in the art conservation world as to which unit is better for monitoring objects of art and cultural heritage. To make matters more complicated, there is no clear relationship between the shock/impact measurement parameters as shown in Table 3.1 and as just described on the one hand, and the stress, strain and eventual damage behavior of objects under shock loading on the other hand. In the (transport) packaging industry, standard packing curves are used to design packaging and damping for specific products/ objects. These packing curves use a combination of g, or in general, acceleration, and velocity. This is based on the concept of the packaging absorbing the energy of a drop or fall. That energy is the kinetic energy, which from basic physics, is related to the square of the velocity as $E_{\text{kin}} = 1/2\ mv^2$ where

E_{kin} is the kinetic energy, m is the mass of the object and v is the velocity at the moment just before the object hits the ground. What g actually stands for is the deceleration of the object as it hits the floor. With good packaging, one wants to make the g value smaller, that is, the object should decelerate slowly and thus absorb the change in velocity (shock) slowly.

A discussion of shock and packing curves is beyond the scope of this book. Suffice it to say that considerable research in the engineering fields continues to be conducted as of the writing of this book to find theoretical and experimental methods to relate units of shock to damage, in particular, experimentally relating so-called shock response spectra (SRS) with damage. A discussion of this work is also beyond the scope of this book.

For the construction industry, the use of velocity is the most common and is the basis for building standards for shock and vibration in most countries. Descriptions of vibrations of trucks and aircraft are also primarily carried out in terms of velocity. As discussed in Section 3.6.3 and in the first paragraph of this section, this is because of the direct relationship between vibration velocity in terms of kinetic energy and the energy to deform a material. It is therefore recommended that velocity be used when dealing with vibrations in conservation. The most important reason is that the combination of velocity and frequency provides the most direct connection to cyclic deformation (see Eq. 3.2) and damage.

3.7 Cyclic Loads: Mixed Shock and Vibration

The reader is probably aware that most cyclic loads which the art conservation world will be confronted with will not be as simple as those described in Sections 3.2 to 3.5. A work of art in a cargo plane will experience a mix of vibration loads, actually starting and ending with the object being transported to and loaded onto the plane, and then continuing with the vibrations of the plane on the taxi and runway, and turbulence and rough weather in the air (see schematic vibration history in Fig. 3.9a). Construction activities themselves are also a source of mixed forms of loading. Pile driving near a museum would be a source of cyclic shock loads (Fig. 3.9b) with the possibility of cumulative damage as a result. One of the most extreme examples of a short-term cyclic load is an earthquake.

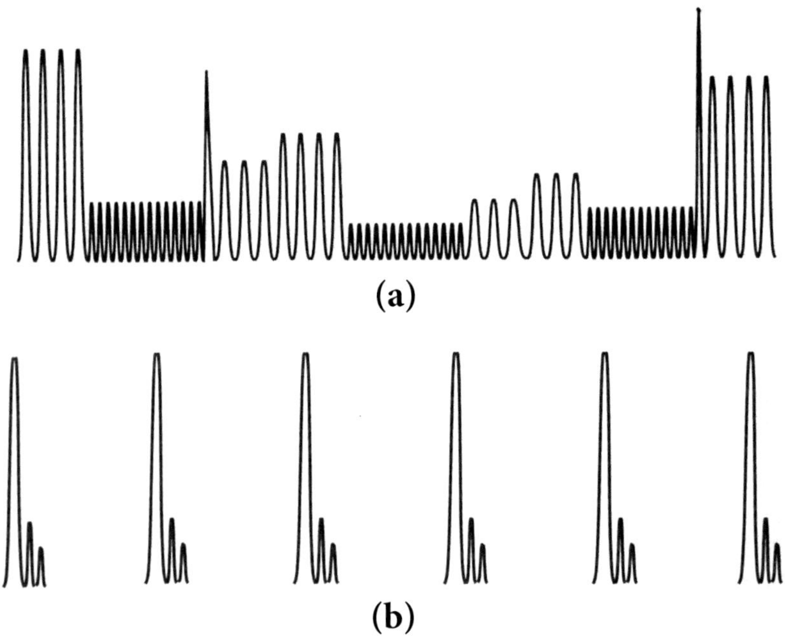

Figure 3.9 Schematic vibration and/or cyclic shock histories which may involve works of art and collections. (a) Transport of a work of art in a cargo plane. (b) Pile driving near a museum.

The analysis of such so-called random vibration signals and relating them to damage and life prediction is quite complex. There are a number of specialized software programs to do this, which are commonly used in the aircraft industry among others. We will briefly describe how such programs work in Sections 5.5–5.6.

Practical exercise 3.3

The purpose of this exercise is to provide an example of how fatigue data in the form of *S–N* diagrams can be used in (preventive) art conservation. It makes use of the fictitious *S–N* diagram shown in Fig. 3.10 for three types of objects of art. The cyclic load levels are given in terms of velocity in mm/s. Notice that the horizontal axis, number of cycles, *N*, is presented using a so-called logarithmic scale. Moving from left to right, each division is 10 times greater than the previous division. The subdivisions are also non-linear as shown in the inset at the lower left of Fig. 3.10. This is the standard way of presenting an *S–N* diagram, which, for many materials, extends to hundreds of millions of cycles. If we were to use a linear scale, the horizontal axis would extend well beyond the walls of the room the reader is sitting in.

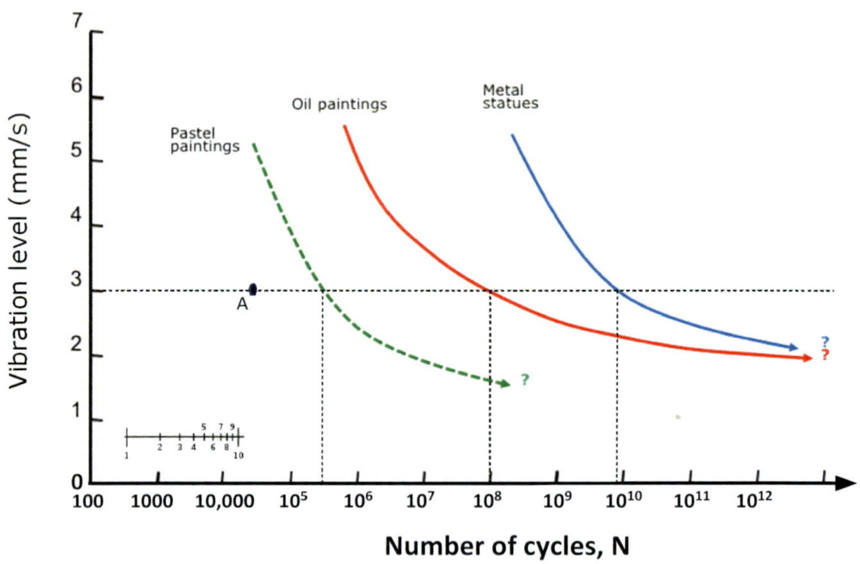

Figure 3.10 Fictitious *S–N* diagram for three classes of objects of art. The horizontal axis is on a logarithmic scale. Values between divisions are divided as shown in the inset at the lower left of the diagram. The dotted black lines are for use with activity A in the Practical exercise.

The activities A through D described below occur occasionally, affecting a collection of paintings and statues, or parts of the collection. For each of the three classes of objects shown in Fig. 3.10 and for each of the activities, answer the following two questions:

1. Can permission be given for the activity to go through as planned?
2. If the answer is "yes" to question 1, how many times could this activity be allowed before one would have to consider conservation treatment?

Use the *S–N* diagram (Fig. 3.10) and the data listed for each activity to help with answering the questions.

The activities are as follows:

A. A rock concert next to the collection during "Night of the Museums." The "beat" frequency is 2 Hz and the concert lasts 3 h. The maximum vibration level at the wall in your museum nearest to the concert will be 3 mm/s.

B. There are plans to route heavy traffic 24 h a day, 7 days a week, next to the collection. This might last up to 20 years. The expected maximum vibration level at the nearest wall is 1.5 mm/s at an average frequency of 50 Hz.

C. A new 10-story wing is planned for the museum. This will involve continuous pile driving for eight (8) work weeks, four (4) hours a day. The pile driver works at one (1) shock every two (2) seconds (0.5 Hz). The expected maximum shock level at the nearest wall is 6.5 mm/s.

D. A request has been made by a partner museum for several of the collection's pastel and oil paintings. It is an eight (8) hour truck drive on an old highway. The truck has top-of-the-line air suspension, and the highway is in fairly good condition. The vibration levels in the transport cases are very low, less than 1 mm/s. However, the highway is made of concrete slabs which over the years have become vertically misaligned. The truck will hit these vertical bumps at an average frequency of one per second (1 Hz). This causes a vibration level of 5 mm/s of the freight bed.

Solution

In order to answer question 1, one needs to first calculate how many cycles, *N*, a given activity will produce. One will then plot the point on the graph relating the calculated number of cycles to the expected cyclic load (vibration) level. If that point is to the left of the *S*–*N* curve for a given class of objects, then the activity is permissible. It that point lies to the right of the curve, the activity is not permissible.

In order to answer question 2, it will be first necessary to find out what the expected life of the class of objects is at the expected cyclic load level. As we discussed using Fig. 3.6, it is the number of cycles where a horizontal line at the expected load level intersects the fatigue curve for that particular class of objects. We then divide this number by the number of load cycles calculated for a particular activity. The number before the decimal point in the result is how many times the activity can be carried out.

We work out the answers for activity A here. For this activity, the input vibration level in the museum is 3 mm/s. The total number of cycles for the rock concert based on the bass "beat" frequency is calculated as follows:

- The "beat" frequency is 2 Hz or 2 cycles per second.
- An hour is 60 minutes × 60 seconds per minute or 3600 seconds long.
- Three hours thus have 3 hours × 3600 seconds per hour = 10,800 seconds.
- The number of "beat" cycles in the rock concert is thus the frequency times the duration of the concert in seconds, or 2 cycles per second × 10,800 seconds = 21,600 cycles.

When the data point is plotted at 3 mm/s cyclic load level (horizontal dashed line) and 21,600 cycles, it will be seen that it lies to the left of all three classes of objects, point A in Fig. 3.10. The answer to question 1 is "yes" for all three classes of objects; the rock concert can go through.

Question 2 is essentially, how many rock concerts can the museum hold before "unacceptable damage" occurs to each of the three classes of objects. In order to answer this question, we first need to determine the number cycles to failure for each class of objects at the expected cyclic load level of 3 mm/s. These values are where the horizontal dashed line at 3 mm/s crosses the three fatigue curves in Fig. 3.10. Using the logarithmic inset in the figure, we find roughly that

$$N_{\text{f\_pastels}} \approx 300{,}000 \text{ cycles,}$$

$$N_{\text{f\_oil paintings}} \approx 100{,}000{,}000 \text{ cycles, and}$$

$$N_{\text{f\_metal statues}} \approx 9{,}000{,}000{,}000 \text{ cycles,}$$

where the "f" in the subscript means failure. If we divide these lives by 21,600 cycles for one rock concert, we see from our fictitious data that

- pastels can survive 13 rock concerts;
- oil paintings can survive 4629 rock concerts;
- metal statues can survive 416,666 rock concerts.

It is noted again that Fig. 3.10 was designed for this exercise using fictitious data. It is only realistic in the sense that pastels are much more sensitive for vibrations than classic oil paintings, and metal statues are the least sensitive.

The other three activities B to D are left as an exercise for the reader. The number of cycles which the reader should get for those activities are as follows:

Activity B: The number of vibrations due to heavy traffic is $\approx 3.2 \times 10^{10}$ (32,000,000,000).

Activity C: The number of shocks of the pile driver is 288,000.

Activity D: The number of bumps the truck experiences is 28,800.

Practical exercise 3.4

If the reader receives strong complaints about performing Practical exercise 3.2, he/she can try the following exercise to get a feel for amplitude and frequency (Section 3.5) as well as for mixed cyclic shock and vibration loads as just discussed, without disturbing those at home. The next time the reader travels with a wheeled suitcase or carry-on, walk across different surfaces with the suitcase (see Fig. 3.11) at a constant speed. Notice the change in frequency and amplitude of the sound one hears while walking across a rough sidewalk surface, and as the suitcase cyclically hits each expansion joint.

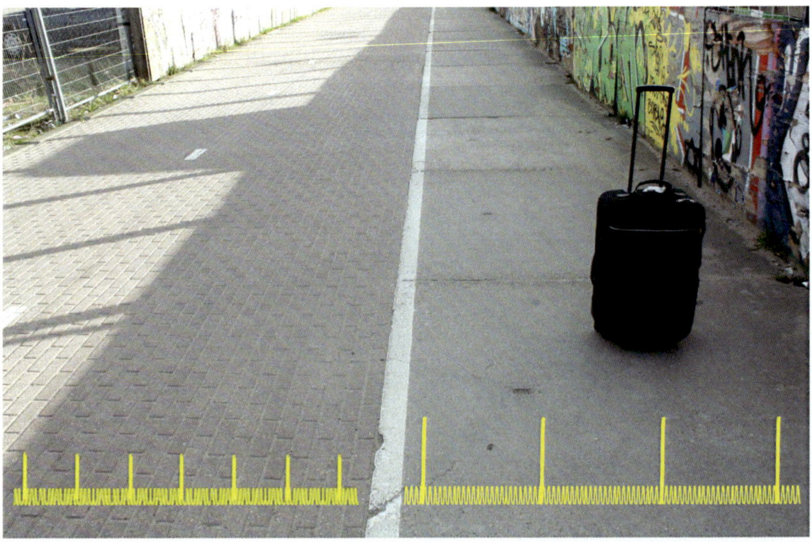

Figure 3.11 Examples of shock due to gaps between the bricks (left half of photograph) or concrete slabs (right half of photograph), mixed with low amplitude high frequency vibration loading due to the roughness of the surface of the bricks and concrete.

In Fig. 3.11a, the frequency of the bumps between the bricks of the bicycle path (to the left) with be somewhat higher than between the concrete slabs of the walkway in Fig. 3.11b. The amplitude of the bumps caused by the bricks of the bicycle path will be slightly less than the bumps caused by the expansion gap between the concrete slabs. If the reader walks faster, all of the frequencies will increase, and vice versa if the reader slows down.

3.8 Cyclic Loads: Resonance

In the previous section, we discussed vibrations as one form of cyclic load. Damage occurs cumulatively depending on the level of the vibrations and the duration (number of cycles). For the higher frequency vibration loads, a dangerous situation can occur when the frequency of the incoming vibrations matches that of the so-called eigenfrequency or **resonant frequency** of the object. At this frequency, the incoming vibrations will be strongly magnified and the objects will vibrate almost uncontrollably.

This type of vibration is what is known as **resonance**. The resonant or eigenfrequency of an object is not a materials property, but is strongly dependent on the geometry and weight of the object as well as the E-modulus of the material. In terms of damage, resonance means that the object is experiencing cyclic loads at a higher level on the *S–N* diagram (Fig. 3.6) than they would at the amplitude of the input frequency.

A consequence of resonance is that one should avoid situations in which the resonant frequency of an incoming vibration source is the same as the eigenfrequency of the source. For example, sensitive objects with eigenfrequencies under 50 Hz which are to be transported by trucks, should be packaged to protect them from the frequencies and eigenfrequencies of the truck riding over the highway. These lie mostly in the range under 25 Hz.

Practical exercise 3.5

A classic example of resonant behavior is the ringing of a wine glass when one rubs the rim at a certain speed with a moistened finger. As the reader may know from personal experience, a specific amount of moisture, rubbing speed and pressure are needed to get the glass to ring. What is happening on a microscopic scale is that the finger is rapidly sticking and slipping on the edge of the glass due to friction. If the frequency of this stick-slip process is the same as that of the eigenfrequency of the glass, one will hear the glass ring at the resonant frequency of the glass. The effect of mass and geometry can be experienced by changing the amount of wine (if the author may be presumptuous) in the glass. The more wine there is in the glass, the lower the tone. There are films on YouTube where professional musicians use this effect on wine glasses precisely filled to obtain accurate musical notes.

Chapter 4

Mechanical Testing

In Chapters 2 and 3, we introduced the basic concepts of the static and dynamic mechanical properties of materials. In this chapter, we will introduce several methods for measuring these properties, including tensile, fatigue and vibration testing. In two exercises, a simple tensile test will be described which the reader can perform at home or in the conservation studio. We will also briefly introduce test methods for measuring special properties such as shear and peel strength, and wetting as related to adhesion strength.

Examples of international standards for the mechanical testing of particular types of materials and load geometries will be introduced for several of the test methods. Many of those are published by ASTM International (formerly known as the American Society for Testing and Materials), ISO (International Organization for Standardization) and CEN (or EN—European Committee for Standardization). These are very often quite similar or even the same, but every country in the industrialized world has its own variations on these standards.

It should be noted that a standard, in the general use of the term, is not necessarily the "right" way to test something. It is

Art Conservation: Mechanical Properties and Testing of Materials
W. (Bill) Wei

ISBN 978-981-5129-63-2 (Hardcover), 978-1-003-61644-3 (eBook)
www.jennystanford.com

a compromise, often developed by commercial competitors who are forced to accept the reality that the consumer wants to be able to compare products tested in the same way. In conservation practice, there will often be applications where the mechanical loading will not be exactly covered by a standard. Standards do, however, provide guidance to make sure that test results are scientifically valid, for example, in how test specimens are designed or loaded. Note that this is true not only in conservation, but also in many areas of applied research and development in industry and academia.

4.1 Mechanical Testing: Objectives and Planning

Before looking at the basics of mechanical testing, it is important to consider what the objective of that testing is and how to plan the testing. This might seem obvious, but the conservation literature is full of testing to determine the "best" material in terms of strength without considering what level of strength is actually required, or whether strength is even the right property for the conservation application.

4.1.1 Objectives

In order to determine the objective(s) for mechanical testing, one must therefore determine what types of mechanical loads/stresses or deformation/strains the object is being subjected to, what the levels of those stresses or strains are, and eventually what other conditions are important. Of equal importance is defining what failure is for the application, and what is acceptable or not in terms of that failure. After all, the ultimate objective of mechanical testing, or any kind of materials testing for that matter, is finding materials that will not "fail" in the application of interest. The development of **failure criteria** is particularly difficult in art conservation because of the subjective nature of the field. This combination of loading conditions and failure criteria determines not only what kind of testing should be conducted, but which properties must be measured, when the test ends, and how the results can be interpreted.

Examples of failure criteria can be found in the practical exercises in Chapter 2. In Practical exercise 2.1, the failure criteria were that the glue had to sustain a particular shear load which was calculated to be 98 N, and that if the joint were to fail again, it had to fail in the same place. Note that in Section 4.7, we will see that the ease of application of the glue may actually be the important failure criterion. In Practical exercise 2.3, a material had to be selected that would be satisfactory for a given load and a given amount of deformation.

As discussed previously, failure criteria do not always involve the strength of the material.

- For many metals, plastics, textiles, and wood, excessive or permanent deformation (yield) is often the failure criterion.
- In Section 4.2.4, we will discuss an application where the E-modulus is the important mechanical property of interest, the mending of cut or torn canvasses.
- When considering the aging of materials, increased brittleness, which is a sharp reduction in failure strain, may be more important than changes in the material strength.
- For cyclic loading, one must consider the combination of stress amplitudes in practice and the duration of the loading until failure. For vibration testing in conservation, this author depends on the conservator to tell him when an object has "failed."

4.1.2 Planning

Once the objectives of the mechanical testing program have been determined, the planning of the tests can begin. Planning involves determining which parameters must be tested and ultimately, how many tests must be carried out. The answers to the questions of how many parameters and how many tests is a balance between "need to know," costs, and time, where the definition of "need to know" is often the critical issue.

As an example, consider a fictive Master's thesis that the author has occasionally been asked to advise on. The objective of the research is to find a glue for repairing a joint. In the literature, the student found five glues that are commonly mentioned for this type of application. They are sometimes used in the as-received condition but are often diluted or otherwise combined with other solvents in various recipes. The student thus wants to try two recipes per glue. Aging is an issue, so testing must also be conducted on specimens aged at various temperatures and relative humidity's for given periods of time. The student comes with the following set of test parameters:

- 5 glues
- 3 recipes including the as-received condition
- 3 temperatures for aging
- 3 relative humidities for aging
- 4 aging times including original state
- 5 repetitions

The number of specimens to be tested in order to account for all combinations of these parameters is 5 × 3 × 3 × 3 × 4 × 5 = 2700 tests. Even if the student planned not to eat or sleep for two or three months, this would be an impossible number of specimens to test for the Master's thesis, let alone for a Ph.D. thesis. One could ask whether testing for four aging times is necessary. Perhaps just the original state and aging for a period simulating 100 years would be sufficient for the objectives of this project. One could also ask the student to look a bit deeper in the literature to find out if all five of the glues are suitable for the application. Perhaps only three of the five would be of interest, and perhaps only one recipe per glue instead of two. Perhaps it would also be sufficient to age at two higher temperatures and relative humidities. While statistics is important in mechanical testing, even in science and industry, cost and time considerations do not always allow for a statistically "ideal" number of tests. The student might therefore want to consider three repetitions instead of five. With these thoughts in mind, the number of required tests is reduced to $3 \times 2 \times 2 \times 2 \times 2 \times 3 = 144$ specimens. This is still a lot, but now the student could consider starting with some initial testing to eliminate possible parameters early on in the testing.

Another consideration in the planning of mechanical testing is what one is looking for in terms of test results. As noted in the example failure criteria above, the student may actually have to consider fracture strain as the failure criterion in addition to or rather than glue strength given that there is an aging component in the research. Part of the planning will also be analyzing the test results, which not only includes the data from the test itself but also examining the specimen fracture surfaces.

Ultimately, the design of the specimen must also be considered. Given that this differs depending on the test, specimen design will be discussed for the specific testing techniques as they are discussed further on in this chapter. For this specific example, the student may be looking at the shear test described in Practical exercise 2.1, or else consider other possibilities such as those described in Section 4.7. If fracture strain is important, the test or specimen design must allow for accurate measurement for what will be a small amount of deformation.

Note that this discussion has only dealt with the mechanical testing part of the project. When developing the ultimate criteria for selecting a particular glue, the student must also consider, for example, the recipes and handling of the glues, and the objectives and planning for the aging of the specimens.

4.2 Tensile Testing

4.2.1 Tensile Testing Equipment

The test used to measure force and deformation, and thus stress and strain as described earlier in Fig. 2.1 is known as a **tensile test**. The equipment used for a tensile test is shown schematically in Fig. 4.1. This is a so-called screw-driven tensile testing system which consists of a stiff frame with a crosshead which is driven up and down using two large motor driven screws. Screw-driven testing systems are preferred over so-called hydraulic systems (see Section 4.4 and Fig. 4.11) for tensile testing. Such systems are much stiffer than hydraulic systems, a property which is especially important if the *E*-modulus (read stiffness) of a material is to be determined.

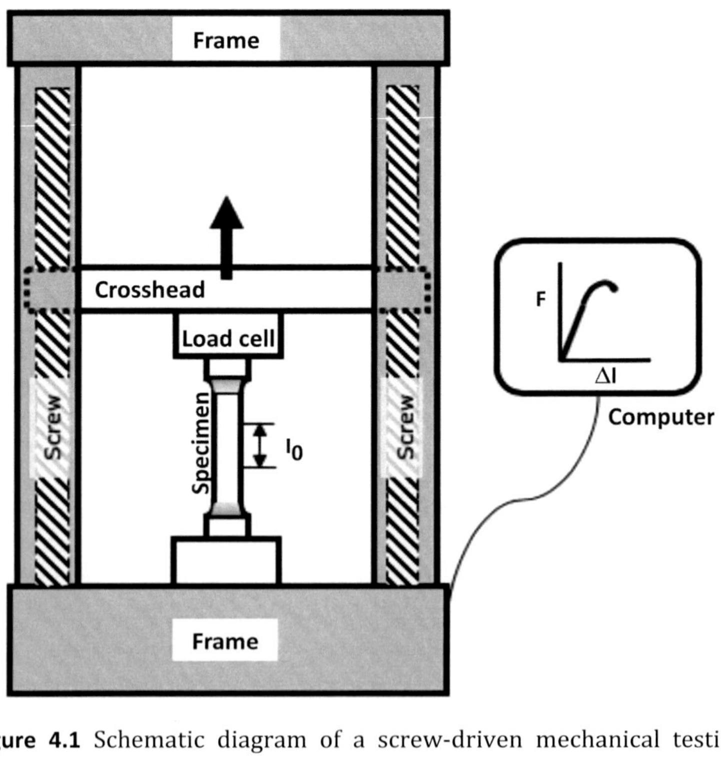

Figure 4.1 Schematic diagram of a screw-driven mechanical testing machine.

A tensile test specimen is connected using grips attached to the fixed frame below, and to a load cell above, which itself is connected to the crosshead (Fig. 4.1). The specimen is loaded in tension by the upward movement of the crosshead. The load is measured by a **load cell** which consists of an electronic circuit which produces a voltage when it is mechanically loaded. Through a calibration procedure, the manufacturer of the load cell can convert the voltage into units of force in Newtons. One can think of a load cell in terms of an old-fashioned scale for weighing fruits and vegetables. Such a scale uses a spring. The change in length of the spring is directly related to weight of the fruits and vegetables placed on it, just as the voltage from the load cell is directly related to the force put on it and the specimen through the movement of the crosshead.

Load cells and test frames are designed for particular loading capacities. The test engineer will select the load cell which is suitable for your test. The load cell must, of course, have sufficient load capacity to be able to carry out your test. However, the load cell capacity must not be too large. You will lose accuracy if you use a high capacity load cell for a weak specimen.

The elongation of the specimen is measured using so-called extensometers (see Fig. 4.2). Such an **extensometer** consists of two arms which are clipped onto the specimen (hence the term clip-on extensometer), and electronic circuitry which produces a voltage which depends on the distance between the arms. The distance between the two arms in the unloaded position is the initial length, l_0, used in Eq. 2.6 for calculating strain. In engineering terms, this initial length is often referred to as the original **gauge length**. The voltage is calibrated to provide accurate measurements of the gauge length, l, and the change in gauge length, Δl. Extensometers are available in various length ranges and accuracies. The test engineer will select the proper extensometer based on the amount of deformation (change in length of the specimen) expected, whether the test is to be carried out until fracture, and the accuracy which is required.

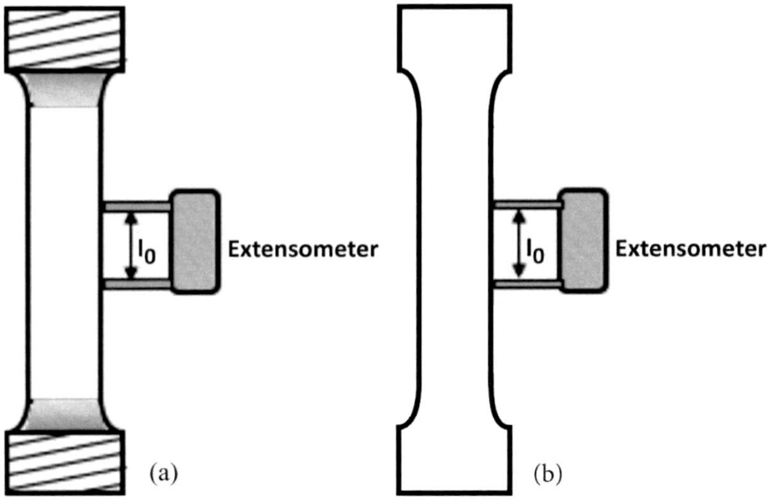

Figure 4.2 Typical forms for dog-bone tensile specimens. (a) Specimen with round cross section. (b) Flat specimen with rectangular cross section.

Tensile test specimens are typically cylindrical or flat (see Figs. 4.2a and 4.2b, respectively). The cylindrical specimens are often threaded at the ends so that they can be screwed into the grips. Flat specimens, as well as some cylindrical specimens are clamped into so-called jaw grips. Tensile test specimens have a so-called dog-bone geometry (see Fig. 4.2) in which the cross-section area is smaller in the middle than at the grips. Recalling that stress is force divided by area, this ensures that the stress in the middle of the specimen is higher than at the grips. The results then reflect the material properties without the influence of the grips (a form of stress concentration).

4.2.2 The Tensile Test

The procedure for a tensile test is fairly straightforward and various aspects thereof were described in Chapter 2. The specimen is mounted in the testing system and the extensometer is attached so that it provides a reading of the original gauge length, l_0. The load or force is then applied by moving the crosshead (Fig. 4.1) upwards at a constant rate. Note that this is a constant elongation or extension rate in, for example, millimeters per minute. A tensile test is thus not controlled by the loading rate, but by the elongation rate.

Modern tensile testing systems are computer controlled and are programmed to automatically measure load and elongation, generating a force vs. elongation diagram as was shown schematically in Fig. 2.8a. The computer can also generate an engineering stress–strain diagram (Fig. 2.8b) if the user provides the original cross-section area, A_0, measured in the middle of the specimen.

There are a number of international standards for tensile testing of various types of materials. Examples published by ASTM include

- ASTM E8/E8M-13 Standard Test Methods for Tension Testing of Metallic Materials;
- ASTM D638 Standard Test Method for Tensile Properties of Plastics;
- ASTM D4964-96(2016) Standard Test Method for Tension and Elongation of Elastic Fabrics;
- ASTM D5035-11(2019) Standard Test Method for Breaking Force and Elongation of Textile Fabrics (Strip Method).

This list is just a brief sampling of many tensile testing standards. While this is all useful information, the practical reality for most conservators is that there is often limited access to tensile testing equipment, more often than not, due to the lack of funding. However, simple experiments can be developed in the studio, and such standards provide useful guidelines for such experiments. Two experiments which the reader can conduct in their studio are described later in this chapter in two practical exercises.

As a final note, the discussion up to this point has focused on tensile testing. The question which may came to mind is how one would test compressive properties of materials.

Compression tests can certainly be conducted using the tensile testing equipment described in Section 4.2.1. The load would be applied by moving the crosshead downwards on the specimen.

Compression testing itself is a bit more complex than that because of the specimen design. One cannot use long specimens as shown, because they may buckle under compression loads. Furthermore, during (plastic) deformation, they will bulge, and final failure is not always a question of the specimen breaking into two pieces. Most conservators of works of art will not need to deal with the compressive properties of materials. For many materials, it has been shown in the literature that the mechanical properties of materials in compression are the same as in tension, except mathematically, they would have a negative sign in front of them. Examples of international standards covering compression tests include

- ASTM E9-19 Standard Test Methods of Compression Testing of Metallic Materials at Room Temperature;
- ASTM D695-15 Standard Test Method for Compressive Properties of Rigid Plastics.

4.2.3 Tensile Specimen Design Considerations

In Section 4.2.1, we noted that load cells and tensile testing systems are available in different load capacities. Theoretically you could look for a system which suits your specimen. However, it is more likely that you will find only one laboratory in your area which has only one testing system. You must then match the size of your specimens to the testing facility rather than the other way around. If you want to test the strength of a material, σ_{UTS}, the force which will be required from the testing system and/or load cell will depend on the cross-section area of your specimen. If your specimen is too large, that is, the cross-section area is too high, then the testing system and/or load cell will not be able to provide enough force to fracture the specimen. You can calculate the maximum cross-section area that the system can handle using, Eq. 2.4, $\sigma = F/A$.

Practical exercise 4.1

A conservator is participating in two different research projects. In one, she/he wants to determine the effect of aging on the strength of Nylon 6 used in a work of art. In another project, the conservator needs to replace a Corten steel load carrying component of an outdoor sculpture. A laboratory near the conservation studio has a mechanical testing system with a load cell that can carry up to 5000 N. What is the maximum cross-section size of the specimens which can be tested?

Solution

We know the maximum force that the testing system is capable of is 5000 N. We can look in the literature on the two materials to estimate what the maximum stress is that we could expect. We then rewrite Eq. 2.4 to solve for the unloaded cross-section area, A_0,

$$A_0 = \frac{F}{\sigma} \tag{4.1}$$

Nylon 6: From the literature, you see that the maximum tensile strength for this material lies somewhere around 90 MPa or N/mm^2. Your aged specimens will probably have a lower strength so you could work with 90 MPa as the maximum stress you might get as a result. Substituting that value and the capacity of the testing system into Eq. 4.1, we get

$$\begin{aligned} A_0 &= \frac{5000\ \text{N}}{90\ \text{N/mm}^2} \\ &= 55.6\ \text{mm}^2 \end{aligned}$$

as the maximum cross-section area the specimen can have. To calculate the diameter of the specimen, you work backwards from the equation for the area of a circle, $A = \pi r^2$, to get the radius, r. Rewriting this equation, you get

$$r^2 = \frac{A}{\pi}, \text{ or}$$

$$r = \sqrt{A/\pi} \tag{4.2}$$

$$= \sqrt{\frac{55.6}{3.14}}$$

$$= 4.2\text{mm.}$$

The diameter of the specimen should not exceed roughly 8 mm (twice r) which, in terms of engineering standards, is a reasonable value.

Corten steel: If one thinks about what we discussed in Chapter 2, one should realize that the goal of the project of replacing a component is that the component should not plastically deform. Thus, the maximum testing force that is actually needed should be that capable of testing up to the yield stress of the material and not the ultimate tensile strength. Some research will be required into how the Corten steel which the artist used was produced. The literature values of the yield stress of Corten steel depend on the conditions it was rolled. If that information cannot be found, one could be dealing with possible yield stresses between 300 MPa and 580 MPa. Let us look at what this means in terms of specimen size.

For a yield stress of 300 MPa, the maximum allowable cross-section area will be

$$A_o = \frac{5000}{300}$$

$$= 16.7 \text{ mm}^2,$$

resulting in a maximum radius of

$$r = \sqrt{\frac{16.7}{3.14}}$$
$$= 2.3\,\text{mm},$$

or a maximum diameter of roughly 4.6 mm. It is small, but such a specimen can be manufactured.

For a yield stress of 580 MPa, the maximum cross-section area will be

$$A_0 = \frac{5000}{580} = 8.6\ \text{mm}^2,$$

resulting in a radius of

$$r = \sqrt{\frac{23.3}{3.14}}$$
$$= 1.7\,\text{mm}$$

This results in a very small diameter specimen of about 3.4 mm which would be difficult to handle. In any case, most standards do not recommend using samples that small for various reasons, unless it can otherwise not be avoided. It would be better if the test laboratory could find a larger load cell or you could find a laboratory with a testing system with a higher load capacity.

Special consideration must also be given to the design of specimens for specific types of materials and how they are manufactured. Not all materials have the same properties in all directions. Wood, for example, is much stronger in the direction parallel to growth than through the tree trunk. A metal or plastic component which has passed through a rolling or extrusion mill will also be stronger in the rolling direction than across it. When conducting tensile testing, one must therefore design the specimen so that the materials properties

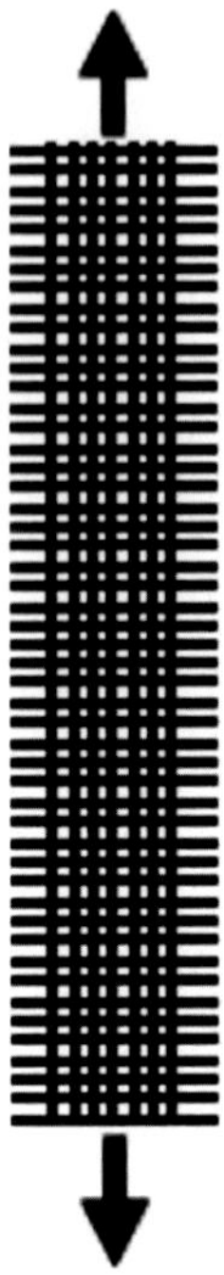

Figure 4.3 Schematic drawing of a textile tensile specimen. Note that standards recommend removing several vertical yarns from each side of the specimen to avoid curling.

are measured in the proper orientation with respect to its manufacturing and how it is loaded in the object.

Many textiles also have different properties in the warp and weft directions, especially since many do not have symmetric weaves. However, they are even more difficult to test because they are not solid materials, but consist of a weave of individual yarns. A dog-bone shaped specimen (Fig. 4.2) cannot be used because the specimen would fail at the transition between the narrow and wider parts of the specimen. One can thus only use rectangular specimens as shown schematically in Fig. 4.3. These are then clamped in the tensile testing machine using jaw grips. This is a form of stress concentration, which will result in some of the specimens failing near the grips. More specimens must therefore be tested than, for example for metals, in order to ensure that enough specimens fail somewhere in the middle away from the grips, so that one obtains statistically valid results. ASTM standard D5035-11 (2019) provides recommended dimensions for such specimens, and also recommends removing a number of vertical yarns on each side to prevent the specimen from curling during the test, a form of the Poisson effect discussed in Section 2.4.

It is noted here that the mechanical properties of textiles are given in different units in the textile industry than they are in other industries. The strength is given in terms of tenacity, which is the maximum force on a force–elongation curve divided by the linear density, that is, Newtons per tex. Tex is the linear density of the textile in grams per 1000 meter.

The reason for the use of such units is because it is difficult to measure the cross-section area of a textile (see Section 5.6). Furthermore, the textile industry deals more with how fabrics "feel" per length, although the industry is, of course, also interested in properties such as tear resistance. However, the research devoted to mechanical properties of textiles, including painting canvasses, generally works with stress and strain as defined in this book (see for example, Section 5.6).

Practical exercise 4.2

In Sections 2.1 and 2.2, we discussed the difference between the terms force and load and the terms stress and strain, and why it is important to distinguish between the two. In Section 2.3, we discussed how one translates force/elongation data from a tensile test into a stress–strain curve. A simple tensile test is now described which can be performed at home or in the conservation studio with readily available materials in order to illustrate the points which were made. (A more practice-oriented version of this test using glued ceramic tiles can be found in reference 1. For that test, the production of specimens will require more effort from the reader.) You will need the following items for this exercise:

- 4 rubber bands (10 mm wide), 2 each of two different lengths (Fig. 4.4a)

- 2 hooks, roughly 5 cm long (Fig. 4.4b)
- 1 plastic bucket with a handle, for example, a container for 1 kg of yogurt (Fig. 4.4c)
- 1 measuring beaker or cup with scale divisions of at least 100 ml (Fig. 4.4de)
- 1–2 l of water
- 1 calipers and/or a fine scale ruler (in millimeters)
- 1 fine tipped permanent marker
- 1 scale for measuring weight (in grams); a digital kitchen scale is sufficient
- 1 steel rod or other structure which allows the entire test assembly to hang freely (refer to Fig. 4.6)

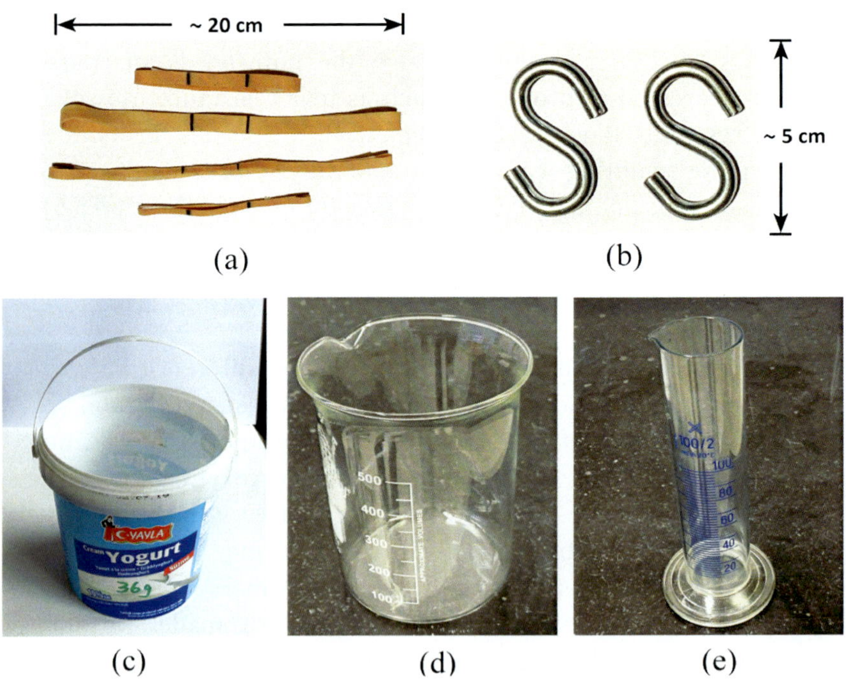

Figure 4.4 Supplies for Practical exercise 4.1 (a) Rubber bands. The top two rubber bands are as-received. The bottom two are the same type but cut in half lengthwise. Sizes are indicative. (b) Hooks. Sizes are indicative. (c) 1 kg plastic yogurt container or similar with handle. (d, e) Measuring beakers with 100 ml division.

Table 4.1 Calculation table for each rubber band in practical exercise 4.1

(see text explanation for what to fill in for the *, # and &)

	1	2	3	4	5	6
	Weight (mass) of water (kg)	Force = weight (mass) × 9.8 (N)	Stress $\sigma = F/A_0$ (N/mm^2)	Gauge length, l (mm)	Elongation (change in gauge length) $\Delta l = l - l_0$ (mm)	Strain $\varepsilon = (l - l_0)/l_0$
1	0	0	0	*	0	0
2	#					
3						
...						
...	&					

*Original distance between marks on the unloaded rubber bands (l_0).

#Weight of hook and bucket.

&Maximum of roughly 500 g for the thin rubber bands, 800 g for the wide rubber bands.

Experimental procedure

1. Construct a table for each rubber band such as that shown in Table 4.1, thus, a total of four tables. This can be done by hand or in some software program such as Microsoft Excel.
2. Cut one of the long rubber bands and one of the short ones in half lengthwise.
3. Take one each of a long wide, long thin, short wide and short thin rubber band (see Fig. 4.4a) for a total of four different rubber bands.
4. For each of those rubber bands, make two straight marks with the fine tipped marker approximately 30 mm apart and perpendicular to the length of the rubber bands as shown in Fig. 4.4a. When doing this, make sure that the rubber bands are flat, but do not use any force which may cause them to stretch. At this point, the distance between the markings does not have to be

precise as long as they are perpendicular to the length of the rubber band.

5. Carefully measure the thickness and width of each rubber band using the calipers and multiply the two values to get the starting cross-section area, A_0. Be careful not to squeeze the rubber band with the calipers while doing the measurements. Save this value. It will be used to calculate the stress in the rubber bands.
6. Weigh one of the hooks and the plastic bucket. Enter this value in Table 4.1 in place of the # sign in Column 1, Row 2, "Weight of water." Make sure to convert the measurement to kilograms.
7. For each rubber band, carefully measure the distance between the two marks using the calipers or fine scale ruler (see Fig. 4.5). Again, make sure the rubber bands are flat, but do not use any lengthwise force to do that. The distance should be measured on either the outside or the inside of both marks. For each rubber band, these are the values of l_0, the original gauge length which will then be entered in Table 4.1 to replace the asterisk, *, in the Column 4, Row 1, "Gauge length *l* (mm)."
8. Hang the hook which was not weighed on a steel rod or beam on which the hook can hang freely without falling off. An example is shown in Fig. 4.6.
9. Take one of the rubber bands, which is now the first specimen and hang it on the hook with the two marks situated in the middle of the length (see again, Fig. 4.6).
10. Hang the hook that was weighed on the rubber band, and then the plastic bucket on the hook (Fig. 4.6). By doing this, one has begun loading the rubber band, which is why their combined weight was entered in Table 4.1 as the first weight entry (Column 1, Row 2).
11. Now measure the distance between the two marks on the rubber band using the calipers or fine scale ruler. Make sure to do this the same way as for l_0, that is, either on the outside of both markings or on the inside. Enter

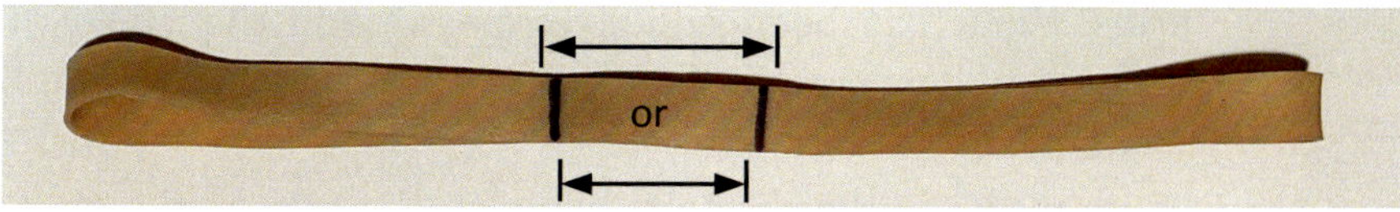

Figure 4.5 Measure the original gauge length, l_0, either on the outside of the two marks or on the inside.

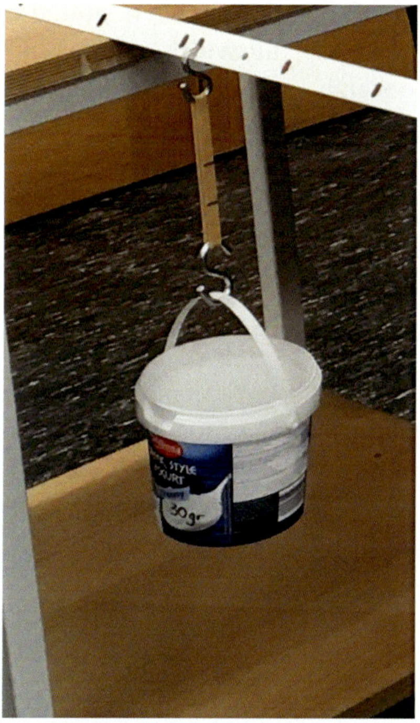

Figure 4.6 Testing assembly hanging on rod or bar which the hooks fit on and can hang freely.

this value in Column 4, Row 2. It is the new gauge length corresponding to the force exerted by the hook and bucket on the rubber band.

12. Using the beaker or measuring cup, pour 100 ml of water into the bucket. Note that 1 ml of water weighs 1 g. 100 g or 0.1 kg has thus been added to the bucket. (Recall that we are using the word "weigh" in layman

terms. What has been added is mass.) Add the 100 g to the previous weight (hook + bucket) and enter the result in Column 1 Row 3.

13. Again, measure the distance between the two marks on the rubber band using the calipers or fine scale ruler and enter the result in Column 4 Row 3.
14. Repeat steps 12 and 13 four more times so that a total of 500 ml of water has been added. If the reader is adventurous, the full-width rubber bands can take a little more weight.
15. The table will now contain seven (or more) values in Columns 1 and 4.
16. Repeat the experiment for the other three rubber bands beginning at step 9.

Analysis instructions

After the experiments have been completed, construct force/ elongation and stress/strain diagrams for each rubber band. In order to do that, it will be necessary to carry out the calculations shown in columns 2/3 and 5/6 in each of the versions of Table 4.1 for each rubber band.

Calculation of force and stress: In order to convert the "weights" of water to force, multiply each entry in Column 1 by 9.8 and enter the results in the same row in Column 2. To convert the resulting force values into stress, divide each result in Column 2 by the load-carrying area and enter the result in Column 3. Be aware that the rubber bands had two "legs" carrying the buckets, so it will be necessary to use twice the value of A_0 which was measured in step 5 of the experimental procedure.

Calculation of elongation and strain: In order to convert the gauge length measurements into strain, first subtract the original gauge length, l_0 (Column 4 Row 1), from the measurements and enter them in Column 5, $\Delta l = l - l_0$. Be sure to always subtract l_0 from each measurement and not the previous measurement from each measurement. To convert the resulting elongation into strain, divide the values in

Column 5 by l_0 (again, Column 4, Row 1), and enter them in Column 6.

Plotting the results: For each rubber band, make two plots of the results of the experiments:

- force (Column 2) on the vertical axis versus elongation (Column 5) on the horizontal axis
- stress (Column 3) on the vertical axis versus strain (Column 6) on the horizontal axis.

This can be done using the software program used to gather the data, or by hand. With a little flexibility in the physical length of the axes, one should be able to plot the force vs. elongation curves for all four rubber bands on one graph, and the stress–strain curves for all four rubber bands on another graph.

Examples of plots for two rubber bands comparable to the top two rubbers bands in Fig. 4.4a with two different lengths and different cross-section areas are shown in Fig. 4.7ab. In Fig. 4.7a, it can be seen that the force–elongation curves separate from each other. The difference is due to a combination of the different lengths and the difference in starting cross-section areas. In Fig. 4.7b, the effect of normalizing the force and elongation values to stress and strain can be seen. The stress–strain curves for both rubber bands are pretty much the same, except for some scatter in the data points which one will always find in such experiments, and as the tests move to higher loads, leaving the linear elastic region for this material. The overlap of the stress–strain curves is due, of course, to the fact that the material is the same for both rubber bands. As we learned in Chapter 2, geometry does not play a role in determining stress–strain behavior.

If one uses similar rubber bands to those shown in Fig. 4.4a, one should get values similar to those in the graphs in Fig. 4.7ab, that is,

- the forces will be less than or equal to roughly 10 N;
- the elongations will be on the order of 10–20 mm;
- the stresses will be less than 1 MPa;
- the strains will be up to 0.5, or in percentage, 50% or so.

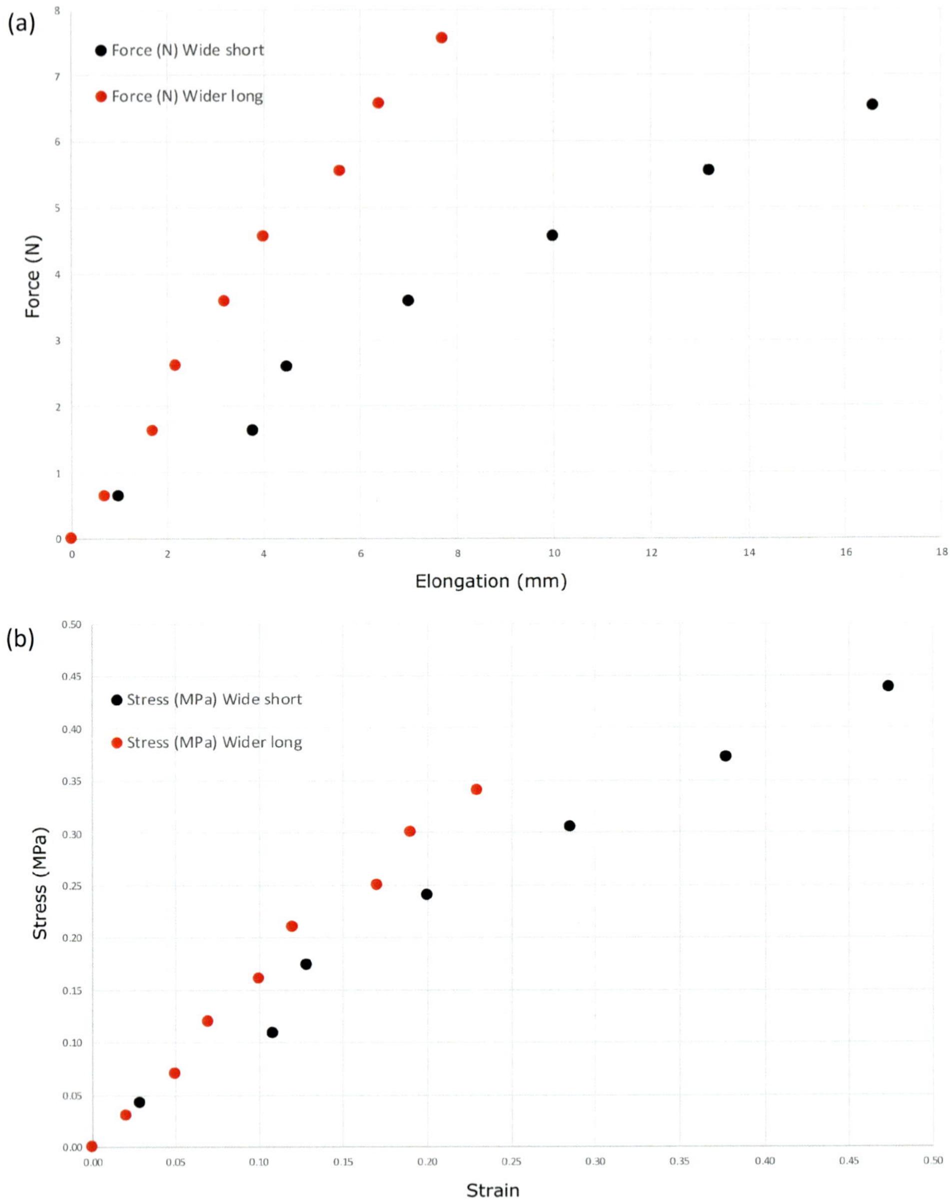

Figure 4.7 Examples of the results of tests on two rubber bands of the same length but of different widths. (a) Force–elongation curves. (b) Stress–strain curves.

There will always be scatter in the results; that is normal. However, if values wildly different from these are obtained, check for the following common calculation errors:

- In Column 1, if the measurements were taken in grams, make sure to convert the values to kilograms by dividing by 1000.
- For Column 3 the width and thickness of the rubber bands were measured in order to get the cross-section load-carrying area. Remember that this needs to be doubled in order to obtain A_0, because the rubber band has two "legs" which are being loaded.
- For Column 5, the change in length is always in reference to the original gauge length, l_0, which was entered in Column 4, Row 1. Δl is thus always calculated by subtracting l_0, from a measurement, and not by subtracting the previous measurement from the current measurement.

Variations on the setup used for testing the rubber bands in Practical exercise 4.2 can be used for conducting simple tests of materials strength using, for example, flat dog-bone specimens, Fig. 4.2b, with holes placed at either end for loading with hooks, see Fig. 4.8. Beware of the specimen dimensions as discussed earlier in Section 4.2.3, and that the holes do not become a stress concentration, see Section 2.6, where the specimens unexpectedly fail.

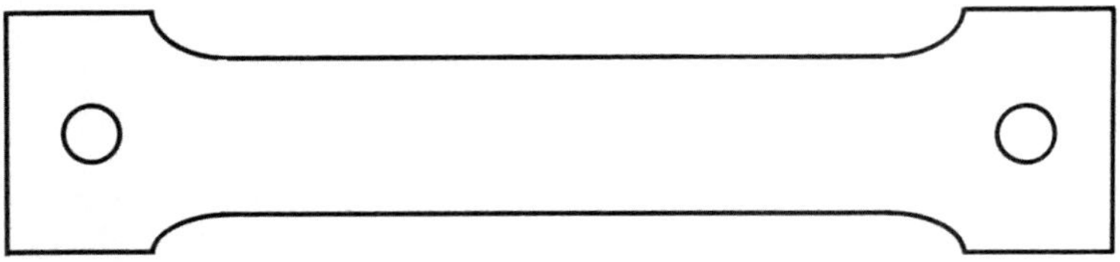

Figure 4.8 Dog-bone specimen with holes for loading with hooks as described for Practical exercise 4.2.

4.2.4 Sewing Cuts and Tears: The *E*-modulus as a Testing Objective

As discussed in Section 4.1.1, the strength of a material is not always the main property of interest when it comes to mechanical properties. An example is a project this author worked on several years ago. (Note that only the details relevant to the discussion in this chapter will be presented.) A large linen canvas painting had been slashed, and it had been decided to sew it back together. One of the main criteria for the success of the restoration was that it should not be visible. Since the linen canvas had a tight weave, this immediately eliminated the option of using the same type of linen thread, simply because there was not enough space between the threads in the weave. The use of the same type of thread would have pushed the weave apart making the sewn area clearly visible.

In order to avoid this problem, the use of polymer surgical threads was the only choice. Surgical threads are available with diameters down to less than one-tenth of a millimeter (one hundred micrometers) so that they easily fit between the yarns in the linen weave. The layperson might think that the selection of the proper type of surgical thread would be based on its strength so that it could hold the two halves of the cut together or its yield stress so that it would not permanently deform. However the situation is more complex than that. Even though the restored painting would be rehung in a museum with climate control, small changes in climate could result in expansion or contraction of both the surgical thread and the original canvas the thread is holding together. Furthermore, in the long term, it would be expected that both materials would age, the linen continuing from the state when it was first painted, and the surgical thread from the moment it becomes part of the painting. This means that both materials must respond in the same way to external factors.

In terms of physical and mechanical properties, the surgical thread must have similar expansion coefficients to the linen, and its mechanical response must also be similar. Considering the latter, as we saw in Section 2.3, the mechanical response of materials is described by the relationship between stress and strain. For linear elastic materials, this relationship is given by Hooke's law, Eq. 2.10, where the *E*-modulus is the parameter that describes the relationship. Thus, in order to find surgical threads with a similar mechanical response as linen, part of the mechanical testing involved in the project had the objective of finding threads that not only were strong enough for the application in terms of yield stress but also had an *E*-modulus close to that of linen.

An example of a set of stress–strain curves from the tensile testing of linen threads and surgical threads is shown in Fig. 4.9. The red data are for linen threads taken from the weft direction in the weave, and the green data are for linen threads taken from the warp direction. Even after an initial period at the beginning of the curve, the warp thread eventually attains an *E*-modulus (slope of the stress–strain curve) which is slightly lower than that in the weft direction.

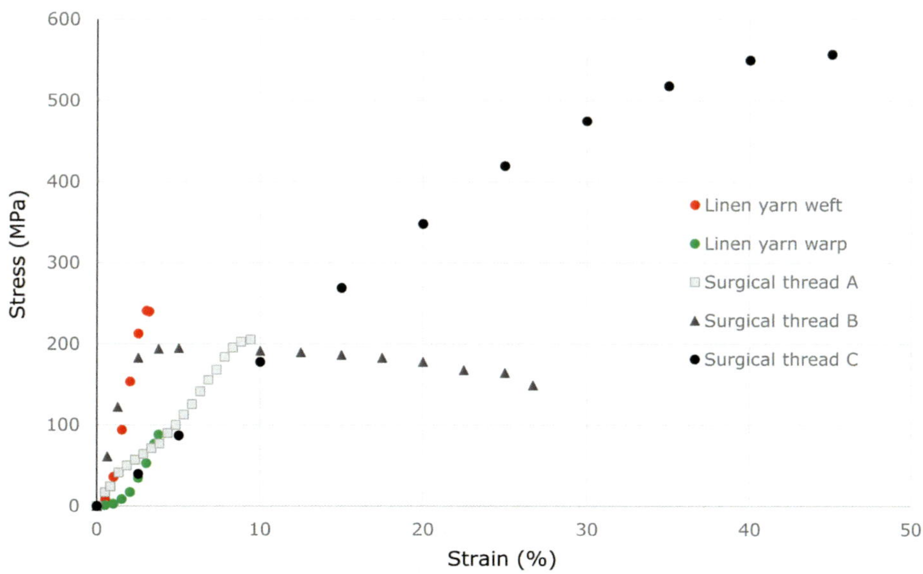

Figure 4.9 Example of stress–strain curves for linen and surgical threads.

Three types of surgical thread were tested and are shown in three gradations of gray. Surgical threads A and C appear to have similar E-moduli, which are slightly lower than for the linen thread in the warp direction. Surgical thread B has a significantly different stress–strain curve as compared with all other materials. It begins with an *E*-modulus similar to that of the linen in the weft direction, but the curve levels off at higher strains.

A complete presentation and analysis of the data from this project is beyond the scope of this section. The point made here is that for this project, a comparison of the geometry and behavior of the sewing materials under loads below the strength of the material was an important factor in the ultimate material selection. Looking at Fig. 4.9 alone, it appears that surgical threads A and C may be possible candidates for sewing the canvas since their E-moduli are at least similar to that of the linen yarn in the warp direction. Surgical thread B has an *E*-modulus similar to the linen yarn in the weft direction but exhibits a clear yield point. Ultimately, a selection was made based on testing not only with threads but also with textiles and mock-ups of repaired canvas materials, as well as perception testing.

4.3 Internal Stresses

In Section 2.7, we discussed the concept of internal stresses. The question often comes up as to whether one can measure internal stresses in painting canvasses or other objects. The answer is no if the measurement has to be performed non-destructively.

In order to measure stress, something has to deform, "move" as it were. In Section 2.3, we developed the relationship between stress and strain in the elastic portion of the stress–strain curve and obtained Eq. 2.10, $\sigma = E \times \varepsilon$. The way the equation reads it appears that one should be able to measure a stress as an independent value. However, recall that this equation is a result of the definition of the elastic modulus,

E. In order to obtain *E*, we had to measure the slope of the stress–strain curve, and that depends on the difference between two strains and the corresponding difference between two stresses (see Fig. 2.8b and Eq. 2.9). In order to come to Eq. 2.9, we conveniently chose one of the stress–strain points to be the start of the stress–strain curve at $\sigma = 0$ and $\varepsilon = 0$. The point is, that in order to measure a stress, there must be a deformation/strain, which implies movement. Furthermore, as we will discuss shortly, we also need to know the *E*-modulus.

The simple description of a load cell in Section 4.2.1 is a case in point. The electric circuit in the load cell is actually deformed during a mechanical test, and the amount of deformation produces the voltage which is then translated into a force. The absolute value of the voltage which comes out of the load cell while it is unloaded is arbitrarily chosen by the manufacturer. The same is true for a strain gauge.

Thus, in order to measure the stress in a stretched painting, one would have to cut a space in the painting to insert a load cell and pull the painting back together to hook it onto the load cell. Pulling the cut section of the painting back together is the deformation needed to reproduce the (internal) tension force that was in the painting. That deformation would be mirrored as a deformation of the electronic circuit of the load cell, which has been calibrated to give the value of the force. To calculate the stress, one would, however, need the *E*-modulus of the canvas/paint combination, which would mean conducting a separate tension test on material which simulates the painting. In short, it is impossible to measure internal stress non-destructively, and this is also true for all industrial applications.

Practical exercise 4.3

Although it is not possible to directly measure stress or strain in a stretched canvas, for conservators and/or artists who stretch painting canvasses, one can get some idea of the strain in a fresh canvas the next time one stretches a canvas. Just before stretching, lay the canvas as flat as possible, being careful not to stretch it in any way. Draw two parallel marks, say, 10 cm apart approximately in the center of the canvas using a sharp pencil, or permanent fine tip pen or felt-tip marker (see Fig. 4.10). The marks do not have to be exactly 10 cm apart. However, the distance between the two marks will need to be measured as accurately as possible in millimeters. Make sure that the measurement is taken either from the inside of one mark to the inside of the other (A_o in Fig. 4.10a), or from the outside to the outside (B_o in Fig. 4.10a). As described for strain gauges in Section 4.2.1, this is your original gauge length for this experiment.

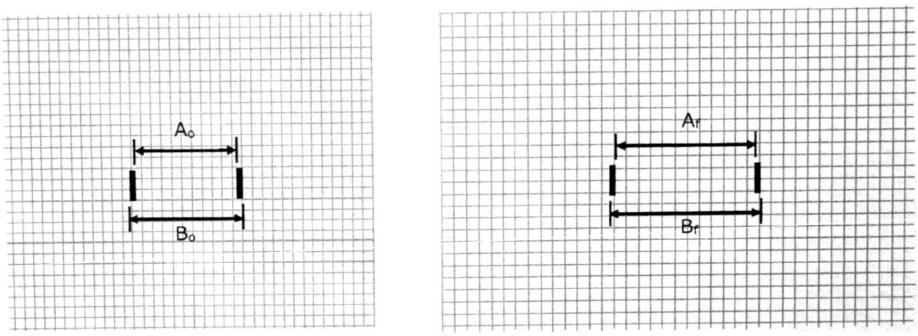

Figure 4.10 Gauge length before and after stretching of a fresh canvas.

Proceed by stretching the canvas as would normally be done. When stretching is completed, measure the distance between the two marks again, the same way as was done before stretching, either on the insides of the mark (A_f in Fig. 4.10b), or on the outsides of the mark (B_f in Fig. 4.10b). This is the deformed length of the gauge length. The strain can then be calculated using Eq. 2.6:

$$\varepsilon = \frac{(A_f - A_o)}{A_o} \text{ or} \tag{4.3}$$

$$\varepsilon = \frac{(B_f - B_o)}{B_o}$$

As discussed just before the beginning of this exercise, if one wants to calculate the internal stress, one will need to multiply this value by the E-modulus, which one will have to either measure oneself or estimate based on values found in the literature.

Note that if the canvas is rectangular (not square), or has an asymmetric weave, the strain that will be measured will depend on the direction it is measured in. One can see that by also placing marks on the canvas in the direction perpendicular to that shown in Fig. 4.10.

———

4.4 Fatigue (Cyclic Load) Testing

Fatigue testing, that is, testing with cyclic loads, can also be conducted using a tensile testing system. Most of the equipment and accessories are similar. However, such testing requires the use of hydraulically driven systems as shown schematically in Fig. 4.11. Screw-driven systems as discussed in Section 4.2.1 and Fig. 4.1 are far too slow to deal with most cyclic load frequencies and loads.

The main features of a hydraulic testing system are similar to those of a screw driven system except that the force is now applied with a hydraulically driven actuator. If one ever uses one of these systems, one will also notice that there is a lot more noise associated with the hydraulic supply unit.

Modern hydraulic testing systems are also computer controlled. Besides controlling the load or deformation amplitude (stress- or strain-controlled testing), the operator can, for example, program the machine to stop after a specified number of cycles instead of testing to fracture, measure and study fatigue crack growth rates (see Section 5.2), or test with varying load amplitude sequences (see the discussion of the Palmgren–Miner rule in Section 5.5).

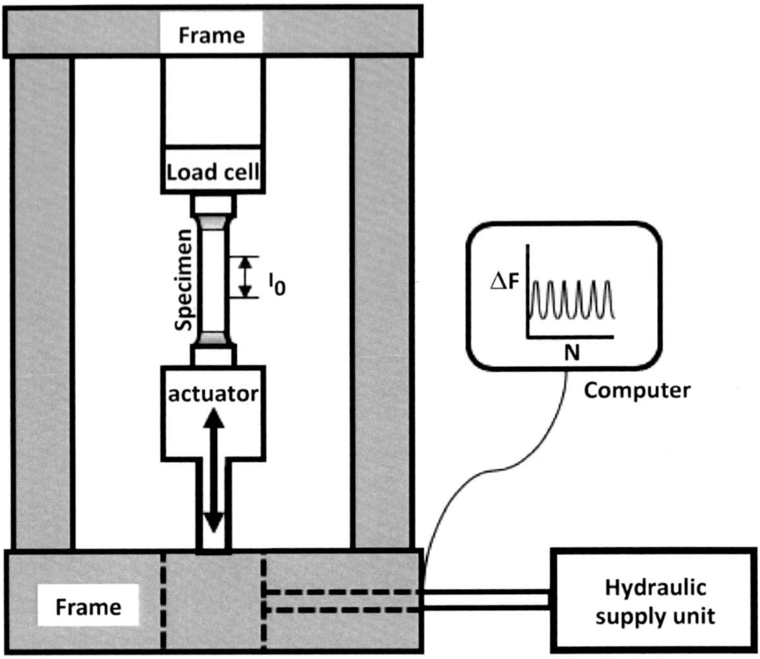

Figure 4.11 Schematic diagram of a hydraulic mechanical testing system. Compare with Fig. 4.1.

Typical standards for fatigue testing include

- ASTM D7791-17 Standard Test Method for Uniaxial Fatigue Properties of Plastics;
- ASTM E466-15 Standard Practice for Conducting Force Controlled Constant Amplitude Axial Fatigue Tests of Metallic Materials;
- ASTM E606/E606M-19e1 Standard Test Method for Strain-Controlled Fatigue Testing.

In Chapters 2 and 3, we discussed sources of mechanical forces which are not purely mechanical. Cyclic changes in temperature, relative humidity or chemical environment can also cause failure, alone or in combination with mechanical loading. Testing with these kinds of loads can be conducted by building a climate chamber around the specimen (see schematic diagram in Fig. 4.12a). The bellows construction shown in the figure ensures that the hydraulic actuator can provide (cyclic) loads if necessary without being limited by the presence of the climate chamber. Temperature cycling can be performed using heating coils/wire (see Fig. 4.12b). This can also be performed at constant load, providing a test where the cyclic load is purely due to thermal cycling. Temperature cycling can also be combined with cyclic mechanical loading in a so-called thermomechanical fatigue test. Most manufacturers of mechanical testing systems offer such chambers, but many testing laboratories have designed their own.

Another common practice in conservation science is to conduct cyclic environmental testing in a climate chamber

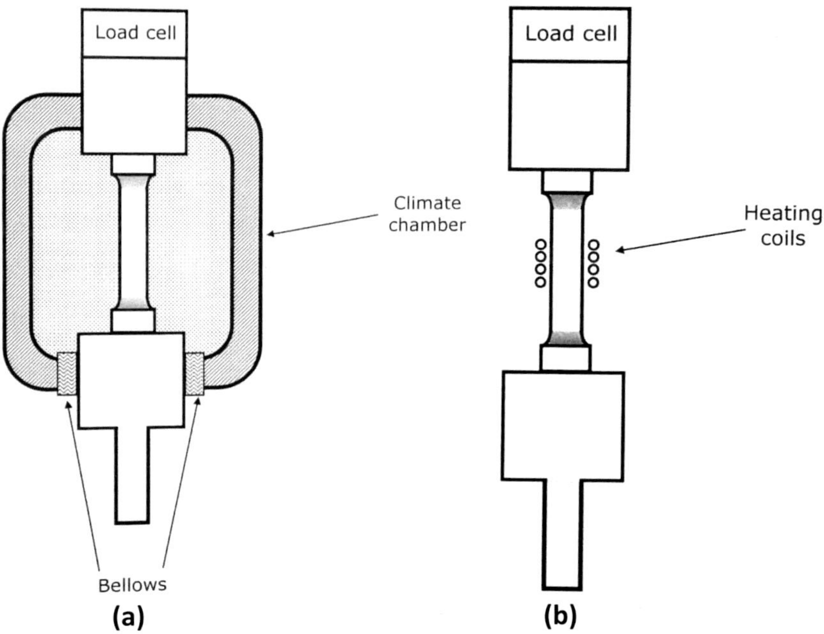

Figure 4.12 Fatigue testing in (a) an environmental chambers or (b) using heating coils. Detail from Fig. 4.11.

ranging from smaller chambers for specimens to walk-in chambers for full-size objects such as large panel paintings or even wooden furniture. Smaller chambers allow for faster changes in temperature and/or humidity so that one can simulate more seasons or years of climate change in a shorter period of time. Larger chambers allow for testing larger objects, but such tests are then concomitantly more expensive.

However, there is a common misconception with the testing of large objects, in particular, with the testing of the effects of cyclic humidity changes. The misconception is that one needs to wait at a certain level of relative humidity until the entire object has reached equilibrium with the environment before moving on to the next (half) cycle. This not only makes the test more expensive, but recalling the shock effects discussed in Section 2.7, this can actually reduce the effect of the difference in moisture content and/or temperature between the surface and internal parts of the object because of the slower build-up of those differences. This can result in non-conservative, longer cyclic life estimates.

4.5 Vibration Testing

Vibration testing of specimens can also be performed with hydraulic systems as described in the previous section. These so-called high-cycle fatigue testing machines are specially designed for operating at frequencies up to around 100 Hertz.

However, much of the vibration testing performed in industry and up until now in the conservation world makes use of industrial vibration tables which are also hydraulically driven such as that shown in Fig. 4.13a. The system shown actually has two testing tables, which produce vibrations in three perpendicular directions (three axes). Any input vibration signal can be produced, from the simplest sine wave as shown in, for example, Fig. 3.3, to complex simulations of truck or aircraft movements. For the conservation field, tests of interest include resonance testing (see Section 3.8),

S–*N* testing (see Section 3.3) at constant vibration levels and frequencies, or simulations of real-life situations and tests of the "wandering" of objects on display shelves. An example of the setup for the testing of the wandering of a ceramic cup on a glass shelf in a museum is shown in Fig. 4.13b.

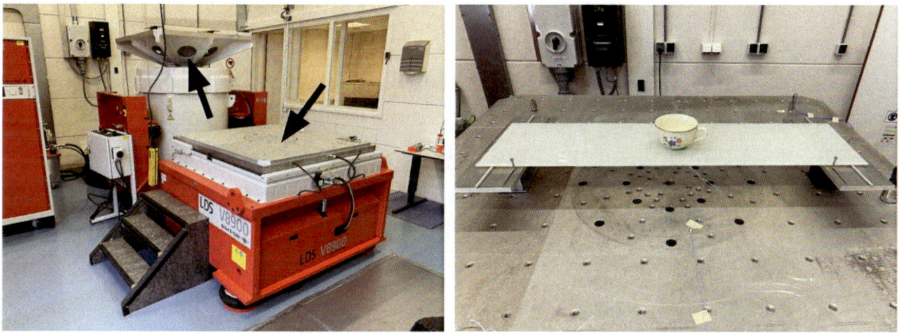

(a) (b)

Figure 4.13 Equipment for conducting practical vibration tests on objects. (a) Typical hydraulically driven system with two vibration tables (arrows). (b) Test setup for the wandering of a ceramic cup on a glass display shelf due to vibrations.

During resonance testing of an object, the system is programmed to continuously increase the input frequency within a certain range at a constant vibration level given in g. For objects of art and cultural heritage, the range of interest is roughly 2 to 100 Hz, with the lower frequencies being the most likely to be associated with damage, depending on the vibration level. (It should be noted that many hydraulic vibration testing systems cannot precisely control frequencies below 5 Hz.)

In order to determine the resonant frequency of an object, one would normally attach a sensor to the test object and compare the vibration level of the object with that of the input vibration to the table (read, for example, floor or wall of a museum or truck). Resonance occurs when the ratio between the vibration level of the object and the input vibration peaks, that is, when

$$Q = \frac{\text{vibration level of object}}{\text{input vibration level to table}} \quad (4.4)$$

reaches a maximum, where the ratio is known as the Q-factor. However, it is very rare that one is allowed to attach a vibration sensor to a museum or historic object, even though the sensors can be very small. The resonant frequency must then be determined by eye. For most objects, it is fairly obvious when they reach the main resonant frequency and this frequency can be noted.

Fatigue testing or testing of the wandering behavior of objects is conducted by placing the object on the vibration table and vibrating it at a programmed vibration level (in terms of velocity, mm/s) and specific frequency typical of the application, or using a simulated load spectrum. The test ends at the number of cycles, N_f, when the object "fails," where failure for a particular object is pre-defined by collections care staff including conservators. For wandering, failure is when the object starts to move, not when it becomes damaged. However, care should, of course, be taken to make sure that the object does not wander off of the test shelf.

The vibration level used for vibration testing could be that which is expected in practice. However, such tests can often be extremely long, especially if one is testing near recommended vibration limits for collections [2, 3]. It may be more useful to test at a higher vibration level and a higher frequency to shorten the test. The life of the object, N_f, which is determined would at least then be conservative. Note that unless there are environmental effects, the frequency used for testing will not affect the number of cycles to failure (fatigue life) for most objects and materials.

One can obtain S–N types of diagrams for given types of objects by testing them at different vibration levels at one frequency and determining the number of cycles to failure.

Vibration levels should be given in vibration velocity (mm/s) (see Section 3.6). For such testing, the resonant frequency is a good one to work with. By changing the input vibration level, one then has a larger range of deformation to work with in order to obtain more levels of S for the S–N diagram. A disadvantage of S–N testing of objects is that either one has to make several "exact" replicas of one object, or one has to test several similar objects, each at a different vibration level to failure. This makes interpretation of the results difficult, either because one has to make general recommendations based on one single replicated object, or based on data of many different objects which will show considerable scatter.

4.6 Shock Testing

As discussed in Sections 3.1 and 3.6, the description of shock loads and the prediction of damage due to shock is quite complex and there is still much ongoing research trying to relate the two. A number of factors determine the behavior of an object under shock/impact loading, among others,

- the properties of the object itself including materials, construction, geometry and weight;
- the orientation in which the object is shock loaded, for example, whether a crate falls flat on one side or on a corner at some angle;
- the properties of the material/object with which the object is impacted.

For industrial applications, there are standards for the shock-testing of shipping containers, products and packaging materials. Examples include among others,

- ASTM D5276-19 Standard Test Method for Drop Test of Loaded Containers by Free Fall;
- ASTM D4003-98(2019)e1 Standard Test Methods for Programmable Horizontal Impact Test for Shipping Containers and Systems.

Some readers may be familiar with more practical drop tests whereby objects simulating works of art are dropped from pre-determined heights and the damage is evaluated. The ASTM standard D5276-19 and other international standards make recommendations on how the objects can be instrumented to measure g at impact and then relate that to damage. In addition, recommendations are made on dealing with the angle of impact. Similarly, there are standards for horizontal impact loads, such as found during the loading of freight onto trucks. A typical instrumented shock sled test bed for horizontal shock testing is shown in Fig. 4.14. The objects to be tested, such as works of art in shipping containers, are placed on the sled, which is then accelerated to the left in the figure.

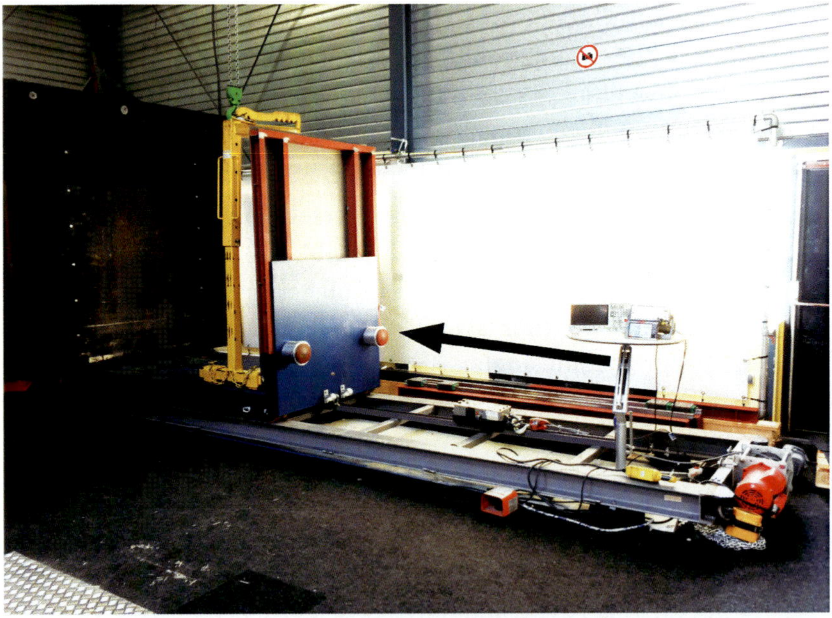

Figure 4.14 Typical instrumented shock sled test bed for testing shipping containers. Motion is indicated by the arrow.

4.7 Testing of Consolidants and Adhesives: Shear and Peel Testing

One question which often comes up in conservation practice is which consolidant or adhesive is better for a particular project. Besides issues of chemistry and appearance, mechanical properties are one of the main concerns.

The testing of the strength of glues and adhesives including consolidants, and the adhesive strength of coatings in general, is one of the more difficult issues in the field of mechanical testing. The mechanical loads on adhesives and the mechanical behavior of adhesives can be quite complex due to the geometry of the loads in practice. This makes the quantitative interpretation of measurements difficult. Adhesives are generally loaded in shear, while consolidants and tapes are meant to "stick" and resist being peeled off of the surface they are attached to. It should thus not be a surprise that we will introduce some concepts of shear and peel testing here as found in the industrial and scientific literature. To conclude, we will also introduce the concept of wetting as an indirect method for measuring adhesion properties.

Shear strength testing

The concept of shear loading was introduced in Chapter 2 during the discussion of stress and strain related to Eqs. 2.4 and 2.5. This type of loading is of particular interest with respect to the use of adhesives in the restoration/repair of various types of objects, and the mounting and hanging of objects such as photographs and textiles. The first exercise to practice using the concept for stress, Practical exercise 2.1, Fig. 2.5, used a typical shear loading on a glue joint which could be encountered in conservation practice.

The testing of the shear properties of materials could be conducted in a tensile testing machine with a specimen as was shown in Fig. 2.5 and is reproduced in another orientation in Fig. 4.15a. This is the basis for a so-called lap shear test, covered, for example, by the standards

- ASTM D1002–10(2019) Standard Test Method for Apparent Shear Strength of Single-Lap-Joint Adhesively Bonded Metal Specimens by Tension Loading (Metal-to-Metal);
- ASTM D3163–01(2014) Standard Test Method for Determining Strength of Adhesively Bonded Rigid Plastic Lap-Shear Joints in Shear by Tension Loading.

A possible problem with this configuration is that it results in twisting of the specimen because the loading line is not

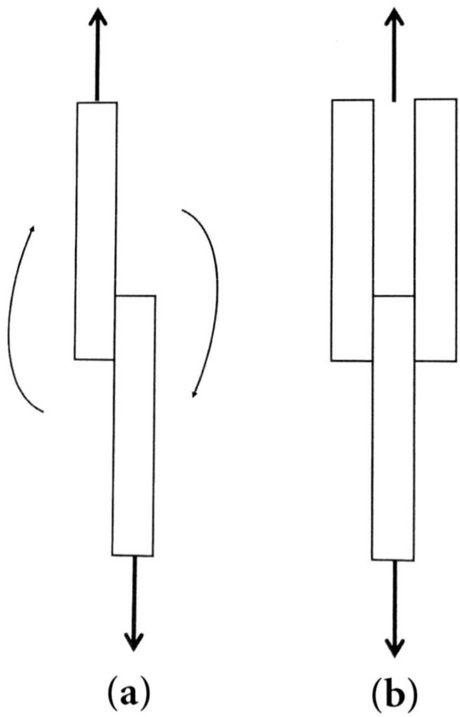

Figure 4.15 Specimens for the testing of adhesive (shear) strength using a tensile testing machine. (a) Lap shear test specimen. The diagram indicates possible issues of twisting. (b) Double lap shear test.

symmetric as it is with a tensile test. This makes the quantitative interpretation of the results difficult. A possible solution is to adhere a second piece of the same material with an adapted grip to apply a symmetric load creating a so-called double lap shear specimen (Fig. 4.15b). Interpretation issues could arise if the two adhered areas do not fail at the same time.

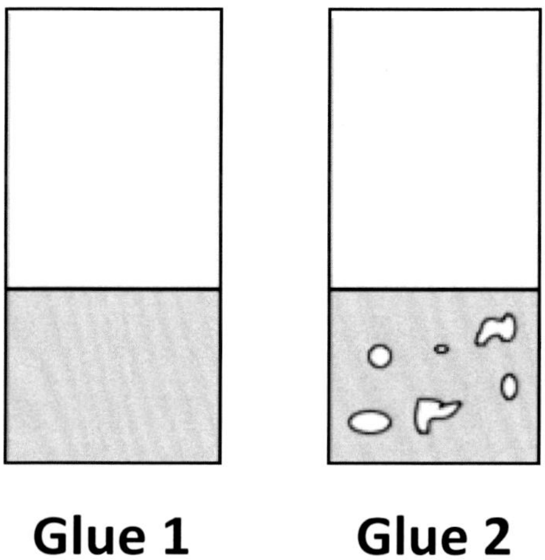

Figure 4.16 Possible fracture surfaces of two glues. Glue 2 could have the same strength as glue 1 if one corrected for the lost load-carrying area of the bubbles. However, glue 2 is harder to apply, making it apparently weaker.

There are two other interpretation issues one should be aware of in testing for conservation applications. If one examines the fracture surfaces of the specimens after testing one may see differences in two glues as shown schematically in Fig. 4.16. Glue 1 may have required more force to shear the joint apart than glue 2, so that the calculated shear strength, $\tau = F/A$ (Eq. 2.5) is higher. However, glue 2 is full of bubbles. If the cross-section area is corrected for the bubbles, the load carrying area, A_2, will actually be smaller than that calculated just by multiplying the height and width of the glued area. One could then possibly calculate the same shear strength for glue 2 as for glue 1. Glue 1 is then not technically stronger than glue 2. Rather, the conclusion is that glue 2 is harder to handle, because bubbles are created during preparation of the glue, or during its application. That would be the more likely reason for rejecting glue 2 for the application.

One should also note that a glue joint actually has three locations where it can fail, see Fig. 4.17, in the glue itself, Fig. 4.17b, in the interface between the glue and the object, Fig. 4.17c, or in the object itself, Fig. 4.17d. Regarding the latter possibility, we noted in Practical exercise 2.1 that one must be sure that a glue is not be too strong for the object materials being restored.

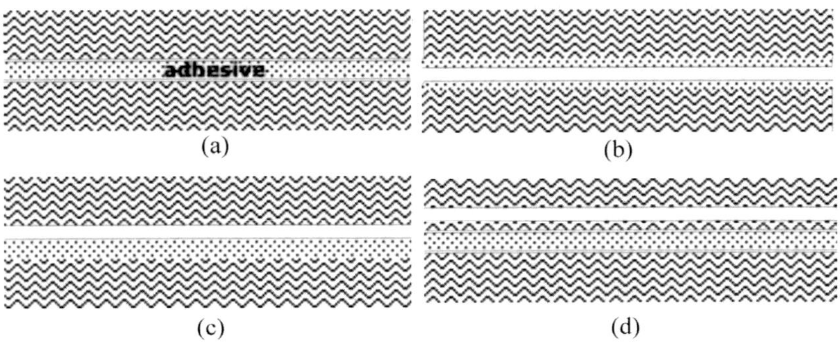

Figure 4.17 Possible failure path in an adhesive joint in an object. (a) Untested joint. (b) In the adhesive itself. (c) In the interface between the adhesive and the object. (d) In the object itself.

Peel testing

The testing of the adhesive strength of consolidants, tapes and other coatings is often conducted based on the way they fail, that is, through peeling. Peeling is caused by a force essentially applied to a line. You can see this when you peel a piece of tape off of a surface (see Fig. 4.18). (As an experiment, this would be more easily visible if you applied the tape to a smooth metal or plastic surface, rather than paper.) As such, there is not a load-carrying area as is the case for the other types of mechanical loading which we have discussed in this book. The concept of stress as force divided by area cannot be used, making comparisons between different tapes, adhesives and coatings in this mode difficult.

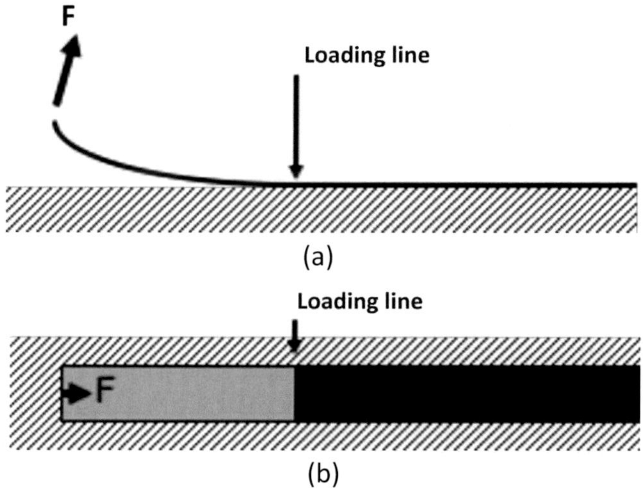

Figure 4.18 Peeling of tape from a surface. (a) Side view. (b) Top view.

There are many standards for testing the peel strength of materials. These include, for example,

- ASTM D903-98(2017) Standard Test Method for Peel or Stripping Strength of Adhesive Bonds;
- ASTM D1876-08(2015)e1 Standard Test Method for Peel Resistance of Adhesives (T-Peel Test);
- EN 28510-1 Adhesives - Peel test for a flexible-bonded-to-rigid test specimen assembly - Part 1: 90° peel;
- ASTM D3359-17 Standard Test Methods for Rating Adhesion by Tape Test;
- ASTM D4541-17 Standard Test Method for Pull-Off Strength of Coatings Using Portable Adhesion Testers.

An example of a test for the adhesive strength of tapes is described by the European standard EN 28510-1 which has a geometry similar to that shown in Fig. 4.18. The tape specimen is pulled upward in a tensile testing system described in Section 4.2.1. The force required to pull 100 mm of the tape from the surface is measured and the results plotted as shown schematically in Fig. 4.19. One can see that the force increases until a short length of the tape comes loose. The force then increases until the next bit of tape comes loose, etc. This slip-stick type of behavior is centered around an average line, the arithmetic mean, which can be calculated by adding up all of the maximum and minimum values in the stick-slip region and dividing that sum by the number of maxima and minima, which in Fig. 4.19 is 22.

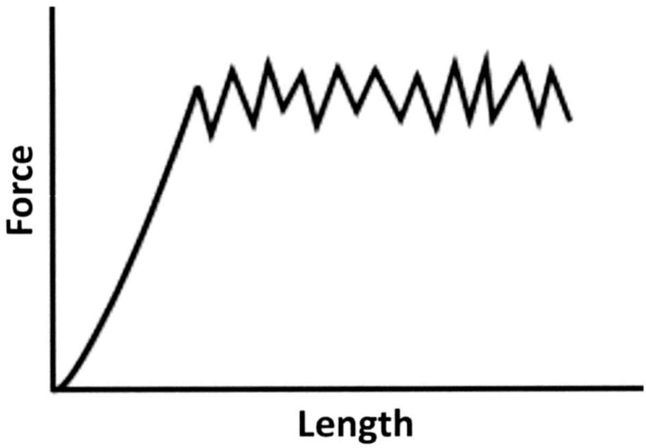

Figure 4.19 Schematic diagram of the results of peel testing conducted according to European standard EN 28510-1.

Wetting

The reader may have noticed that the preceding discussion of shear and peel testing involves the use of mechanical testing equipment. Some testing standards also recommend special specimens, specimen holders and other kinds of mechanical loading systems. Such equipment will often not be available or affordable to conservators. Furthermore, the direct applicability of the results of such testing for consolidants which will only be used in small amounts on, for example, a painting, may be questionable.

One property of materials in liquid form that may be useful to provide a rough indication of their adhesive strength is their ability to wet a surface. If a drop of liquid is carefully placed on the surface of interest, it will contact the surface at a particular angle, the so-called contact angle. This is the angle between the surface and the tangent to the drop where it contacts the surface (Fig. 4.20). If the contact angle is less than 90° (Fig. 4.20a), the liquid will wet the surface, if it is greater, it will not (Fig. 4.20b).

A good example of poor wetting is a car which has just been waxed. When a raindrop falls on the waxed surface, it beads up and rolls easily off. The contact angle is close to 180°. On the other hand, on a dirty car, raindrops immediately spread out, connecting in a water film. Each raindrop has a contact angle close to 0° and it wets the surface of the car. You must then use a towel with some force, minimal but not zero, to dry up the water film.

If one wants to compare the adhesion properties of several possible adhesives or consolidants, for example, consolidants

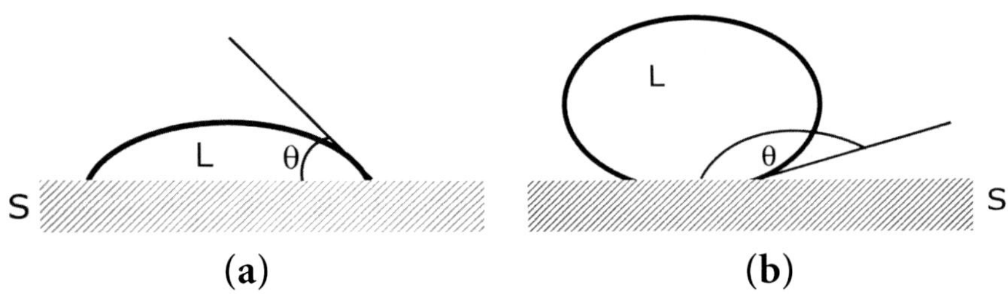

Figure 4.20 Contact angle between a drop of liquid (L) and a solid surface (S). (a) Wetting if the contact angle, θ, is less than 90°. (b) Poor wetting if the contact angle, θ, is greater than 90°.

for treating a painting, one can conduct a relatively simple test as shown in Fig. 4.21. Place a drop of each consolidant on a paint surface which is the same as that on which the consolidant will be used. Using a high-resolution consumer camera, take a picture of the droplet, and then measure the contact angle on the resulting image. The results will give an indication as to which consolidant wets the paint surface better. Consolidants with high wetting angles will probably not adhere well to the paint surface when dry. Consolidants which have low wetting angles are more likely to adhere well. If several consolidants have contact angles less than 90°, it is possible that the consolidant with the lowest contact angle will adhere better, though this is not always the case. Other tests and consultation of the literature and experts are then required to help make the final selection.

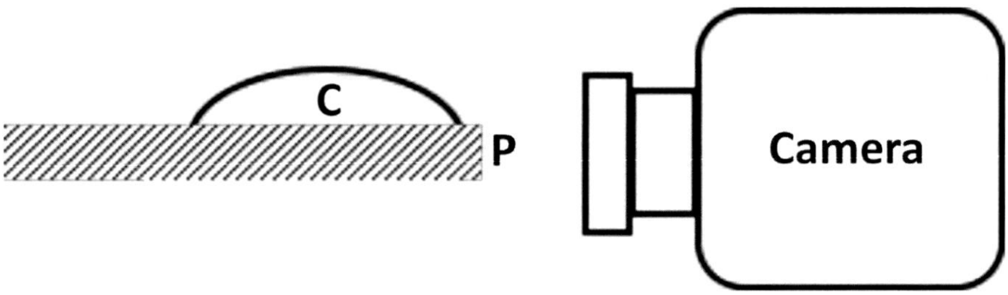

Figure 4.21 Measurement of the contact angle between a consolidant (C) in liquid form and a paint layer (P) using a good high-resolution consumer camera.

4.8 Monitoring

In this section, several methods for the *in situ* monitoring of the mechanical behavior of objects will be introduced. Although monitoring is not mechanical testing *per se* as meant in this chapter, it can be useful for providing an indication of strain levels for designing mechanical testing in the laboratory or conservation studio, and determining the mechanical behavior of materials and objects in the field when mechanical testing is not possible.

Before going into the details of the monitoring methods, it is important to note three points. The first is that what is being monitored are changes in deformation or strain, not stress. As was discussed in Section 4.3 on internal stresses, it is not possible to directly measure stresses in an object. In everyday language, engineers will talk about measuring stresses *in situ*, for example, in a building. However, what they are actually doing is measuring the deformation or strain.[1] What one is doing when monitoring the mechanical behavior of objects *in situ* is thus measuring changes in the deformation/strain. If the material is behaving elastically, one can then estimate the stresses using Hooke's law, Eq. 2.10 in Section 2.3.

The second thing to note is that all common methods for the *in situ* monitoring of strain require the firm, direct application of the measurement sensors to the object using some sort of adhesive or by mechanical means. For many, or perhaps better said, for most works of art and cultural heritage, this may limit what and how one is allowed to monitor, if at all. In fact, the following techniques are most often used for monitoring deformation and cracking in historic buildings, museums, and large statues, and for monitoring incoming vibrations near the objects of concern in museums or transit cases.

Finally, it is important to note that monitoring is good for obtaining information about mechanical behavior in practice. However, monitoring does not itself protect works of art or cultural heritage. In order to do that, an action plan is also required to define limits beyond which mechanical properties should not exceed, and actions and consequences if they do.

[1]Also recall from Sections 4.2.1 and 4.3, that a load cell in a tensile testing machine is actually measuring the deformation of its electric circuitry, producing a voltage which is then calibrated into force.

4.8.1 Strain Gauges

The measurement of the deformation/strain of parts of an object is often conducted using so-called strain gauges, a typical form of which is shown in Fig. 4.22. A strain gauge consists of a specially designed pattern or network of electrical resistors. The deformation of this pattern/network results in a voltage that is proportional to the strain, where the initial length, l_0, as described in Section 2.2 is defined by the length of the strain gauge.

Strain gauges are mounted in areas of high stress, for example, on top of the arm of the statue next to the shoulder, see Fig. 4.22. The goal of monitoring at such locations is to ensure that the strain does not exceed, for example, 0.002 (yield) for a metal object, or the failure strain of a ceramic or stone stature.

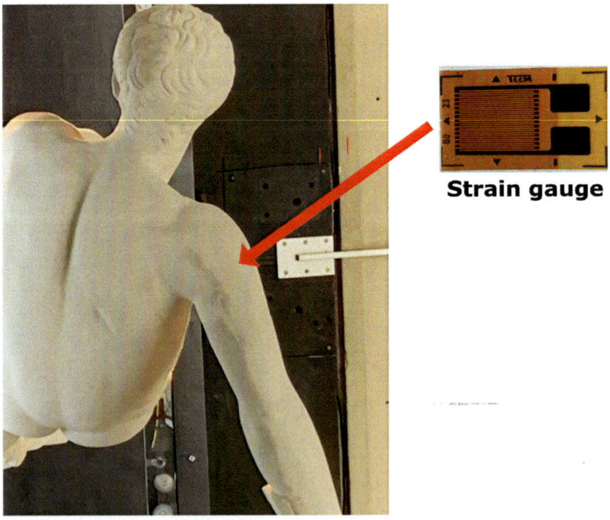

Figure 4.22 Typical strain gauge that can be mounted using an adhesive on an object at an area of high stress such as the top of the arm next to the shoulder of the statue (photograph of strain gauge by Pleriche, own work, CC BY-SA 4.0, https://commons.wikimedia.org/w/index.php?curid=61871859; photograph of the statue by this author).

4.8.2 Crack Opening Displacement

Cracking in, for example, load-bearing structures such as the balcony shown in Fig. 3.5, wall paintings on load-bearing walls, or in large works of art can often be a cause for great concern. The concern lies in the possibility of further crack growth in environments with large cyclic changes in temperature and/or humidity, and in structures that undergo considerable cyclic loading, leading to building and visitor safety issues.

In order to determine if a crack is still active[2] under such cyclic loads, it would be useful to measure the growth (or extension) of the crack at the tip of the crack. However, this is very difficult, if not impossible, because the crack tip is very fine, and thus hard to find or is hidden behind a layer of paint, loose plaster, or dirt films.

A common way for monitoring crack growth (or extension) is thus to measure the change in separation of the two halves of the crack at some location along the crack. This is known as measuring the crack opening displacement, abbreviated COD. As a crack grows under cyclic loading, the separation or opening between the two halves of the crack at a specific location along the crack will also grow. The COD can be measured simply, for example, by placing a transparent millimeter scale over the crack and noting the crack opening at regular time intervals, see Fig. 4.23a. In terms of measuring equipment, this is an inexpensive way of measuring COD, where the resolution can be on the order of fractions of a millimeter.

For higher accuracy, one can make use of electronic extensometers such as so-called LVDTs (linear variable differential transformers). Simply put, an LVDT consists of a magnetic core that moves inside a set of coils (transformer), see schematic diagram in Fig. 4.23b. The core is permanently affixed to one side of the crack, and the coils to the other. When the crack opening changes, the magnetic core moves

[2]Note that the presence of cracks under cyclic mechanical or climate loading does not always mean that they are active, see for example, Section 5.2.2 Craquelure and fracture mechanics, further on in this book.

relative to the coils, red arrow in Fig. 4.23b, producing a voltage that is directly related to the relative motion of the two halves of the crack. The interpretation of changes in COD with respect to crack growth falls in the realm of the engineering field known as fracture mechanics, which is briefly introduced in Section 5.2. The theoretical details of the interpretation of COD data are beyond the scope of this book.

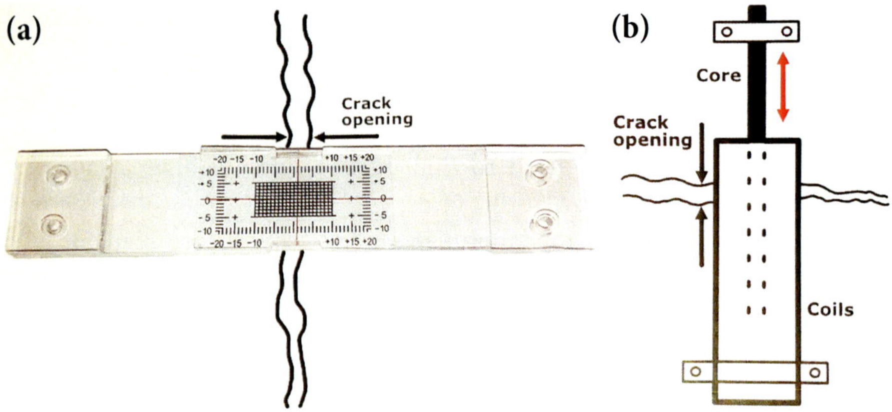

Figure 4.23 Measurement of crack growth by measuring the crack opening displacement using (a) (sub)millimeter scale, (b) LVDT extensometer—as the crack opening changes, the core moves (red arrow).

4.8.3 Vibrations and Shock

Strain gauges and COD monitors can be used for many situations where cyclic loads are present, ranging, for example, from seasonal climate changes to pedestrian traffic on historic balconies. However, there are limits to the cyclic load frequency that can be measured with such sensors. In particular, they cannot respond fast enough to vibrations or shock loads.

For monitoring vibrations and shock, sensors known as accelerometers are used. Such sensors consist of special materials, often the so-called piezoelectric materials that produce an electrical signal when they are accelerated. The signal is related to acceleration and can be used as such, or preferably, converted mathematically to velocity or displacement (deformation). The conversion depends on the frequencies of the vibration or shock loads. Accelerometers come in various sizes and sensitivities. Some of the most sensitive, but also largest, are the so-called vibrometers or seismometers, which are commonly used, as the latter name implies, for detecting earthquakes. Sensors that are of interest for conservation applications can be as small as several millimeters in diameter, but still with high sensitivity, more than sufficient for monitoring vibrations of interest with respect to works of art and cultural heritage. They can then be relatively easily mounted almost unseen in museums or on the inside of transport cases.

An example of a vibration measurement system used to measure vibrations in a museum storage facility during a rock concert in the museum is shown in Fig. 4.24. In Figs. 4.24ab one sees an accelerometer, A, which measures in one direction (1D) being used to measure the vibration of the floor in the vertical direction. In Figs. 4.24ac, a slightly larger accelerometer, B, can be seen which measures the vibrations in a rack in three directions (3D) perpendicular to each other. The datalogger is also shown, C in Fig. 4.24a. In such situations, the accelerometers are mounted using thin double-sided tape, or a thin layer of beeswax. Again, this means that they can generally not be directly attached to objects unless it is possible to clean off the traces of the tape adhesive or beeswax without damaging the object. However, vibration monitoring is most often conducted in particular rooms/galleries or in transit, therefore, it is sufficient to mount the accelerometers on the walls or floor near the objects of concern, or at some pre-defined location nearest to the source of the vibrations in the gallery of concern in a museum. In transit, they can be mounted in and/or on the transport crate or, for example, on the transport frame of a painting.

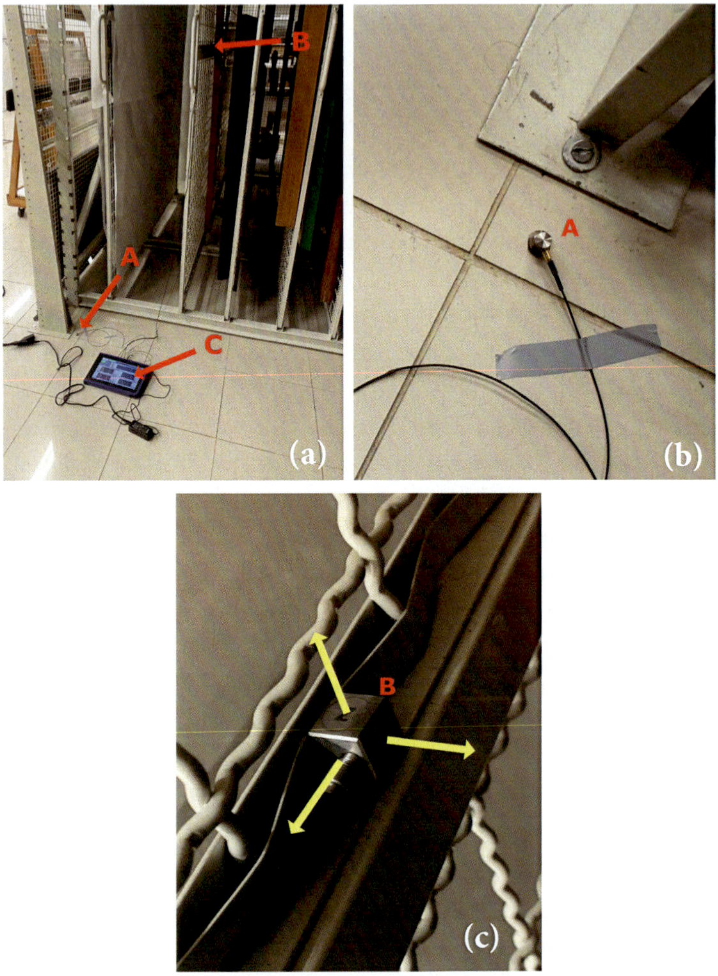

Figure 4.24 Location of vibration measurement sensors in a storage depot. A: 1D sensor; B: 3D sensor; C: datalogger. (a) Overview; (b) 1D sensor on the floor next to a rack system frame (measures perpendicular to the floor); (c) 3D sensor on a moveable rack (yellow arrows indicate three measuring directions perpendicular to each other).

Note that, unfortunately, the term "accelerometer" has contributed to much confusion in the cultural heritage field on what units should be used to describe vibrations. As discussed in Section 3.6, vibration levels should be given in units of velocity, *e.g.*, mm/s, while shock loads are generally given in units of acceleration, *e.g.*, mm/s^2 or g.

References

1. Wei, W. (2014). Teaching the concepts of the mechanical properties of materials in conservation, in *CeRoArt*, ed. W. Baatz. https://journals.openedition.org/ceroart/4215 (accessed 5 January 2021).
2. Wei, W., Sauvage, L., and Wölk, J. (2014). Baseline limits for allowable vibrations for objects, *Proc. ICOM-CC 16th Triennial Meeting*, Melbourne, Australia, 15–19 September 2014, Paper 1516, (International Council of Museums, Paris).
3. Wei, W., Watts, S. Seddon, T., and Crombie, D. (2018). Protecting museum collections from vibrations due to construction: ibration statistics, limits, flexibility and cooperation, *Studies in Conservation* **63** S1 pp. S293–S300.

Chapter 5

Advanced Concepts

In the first three chapters of this book, we introduced the basic concepts of stress and strain in materials and objects, and the difference between static and dynamic loads. We followed this with a chapter on the basics of mechanical testing. The material in the first four chapters was based on relatively simple loads which are applied in one direction, either tensile, compressive or shear. We showed a number of examples in conservation where this situation applies, and that one can obtain initial answers to many questions concerning the behavior of objects and materials in conservation based on simple calculations.

As the reader is well aware, many situations in real-life involving mechanical loads are much more complex. Solutions to mechanical problems become more difficult when there are several stresses acting in different directions, or when the load is changing randomly with time. In this chapter, we will introduce several new variations on the types of mechanical loading we have already seen. These include creep, stress relaxation, the friction and wear of materials, the combined effects of mechanical loads and the environment, and crack

Art Conservation: Mechanical Properties and Testing of Materials
W. (Bill) Wei

ISBN 978-981-5129-63-2 (Hardcover), 978-1-003-61644-3 (eBook)
www.jennystanford.com

propagation. Finally, we will describe the basics of several mathematical models for helping to find solutions to protect objects and collections from complex mechanical loads, where mechanical testing may be too difficult to perform.

5.1 Stress Relaxation and Creep

In Chapter 2, we discussed the concepts of stress and strain under static loading. Under a given stress, the material is strained a certain amount in a very short time, and both then remain constant with time. However, for certain materials under certain conditions, the stress or strain will change with time after this initial loading. These changes are accompanied by permanent deformation of the material and can eventually lead to failure. The two processes which lead to this time dependency are described briefly in the following paragraphs.

5.1.1 Stress Relaxation

When certain types of materials are deformed to a *given strain*, the resulting stress in the material reduces over time in a process known as **stress relaxation**. This is shown schematically in Fig. 5.1. The relaxation is caused by local deformation on a microscale. In many polymers and plastics, such deformation processes can occur at room temperature, see the following practical exercise. However, they often also occur near the glass transition temperature of the materials. In any case, they involve viscous flow (flow of polymer chains past each other), disentangling of polymer chains and the breaking of chains (scission). As this local deformation occurs, local stresses are released. Since the original macrostrain is constant, this means that the macrostress will be reduced. It also means that if the macrostress is removed, the microstrain which has built up will remain, resulting in permanent deformation of the material or object.

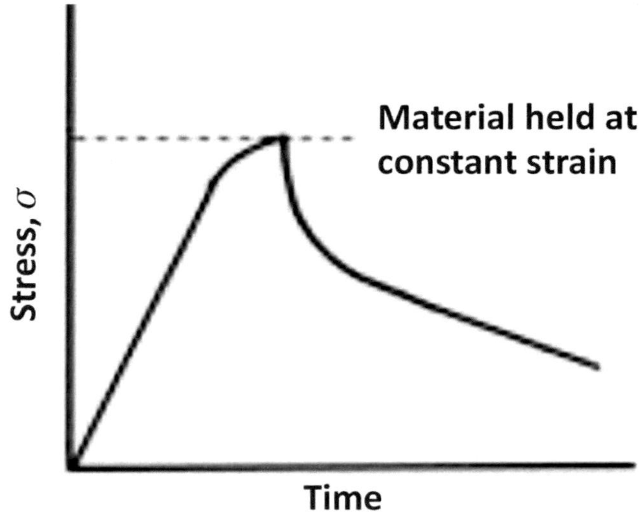

Figure 5.1 Schematic diagram demonstrating the concept of stress relaxation. The stress relaxes under constant strain.

A classic example of stress relaxation can be found in paintings conservation, the loss of tension in a painting canvas with time. When a canvas is stretched, the stress causes a straightening of the threads in the canvas weave, and the threads themselves also start to untwist. This microstrain also leads to local reduction in the stress and thus in the macrostress. However, the weave structure will no longer be able to return to its original form, so this deformation is permanent. You will notice this when trying to re-stretch a painting. If you use the same force to stretch the painting as the original force, the crease in the canvas originally caused by the stretcher will go beyond the stretcher. For very large paintings, the original tensile stress in the canvas will disappear, and the canvas may begin to sag and eventually bulge out under its own weight. Restretching brings a new tensile stress state into the canvas, but not the same as that which was brought in by the original stretching of the canvas.

Practical exercise 5.1

Stress relaxation can be observed at room temperature by stretching a rubber band tightly around two solidly fixed objects such as the handles of a small metal chest of drawers (see Fig. 5.2). Right after the rubber bands have been stretched, record the tone of the two vertical parts of the rubber band as they are plucked. In the next day or two, the tone will noticeably drop. This is due to the relaxation of the stress in the rubber band. The lower the tension, the lower the tone. Note that the strain in the rubber band remains the same, that is, in Fig. 5.2, the distance between the two drawer handles. After a week or two, the tone will slowly reach a steady tone. This experiment is analogous to what happens to a violin string made with natural gut, which is tuned and then allowed to sit for days or weeks without playing.

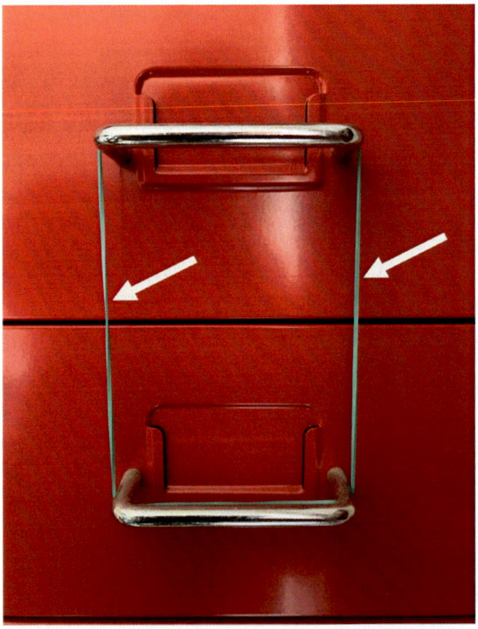

Figure 5.2 Experimental setup for observing stress relaxation of a rubber band (arrows) stretched around two drawer handles.

5.1.2 Creep

When certain types of materials are loaded to a *given stress level*, they will initially deform to the strain level corresponding to that stress as given in their normal stress–strain diagram as described in Chapter 2. This can be seen at the extreme left of the schematic diagram of strain versus time shown in Fig. 5.3. However, unlike the situation with static loading, such materials will continue to deform over time at constant stress in a process known as **creep**. This begins immediately after the initial loading in what is known as stage I or primary creep. At some point in time, the strain will increase at a constant rate known as the steady-state creep rate. This occurs in stage II creep in Fig. 5.3 and is also known as secondary creep. At high strains, the specimen will start to neck and the creep rate will increase until failure. In Fig. 5.3, this is shown as stage III creep and is also known as tertiary creep. In all stages, if the specimen is unloaded, the strain which has occurred during the creep process remains; it is a permanent strain.

The continuous increase in strain during creep is due to many of the same microstructural changes which occur

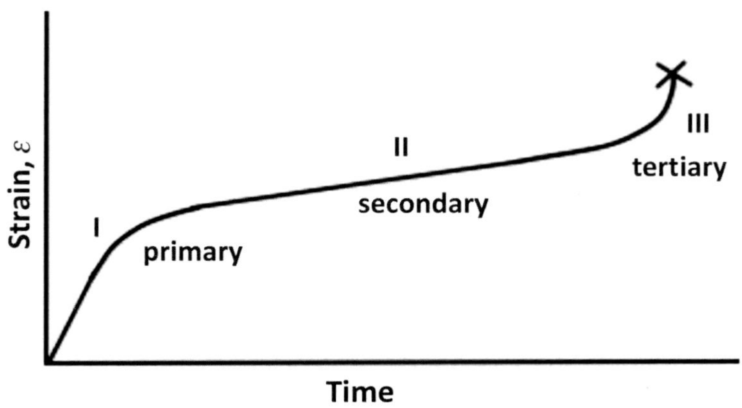

Figure 5.3 Schematic diagram showing the three stages of creep.

during stress relaxation. This often leads to confusion between the two. Remember that creep is a continuous increase in strain under constant stress, whereas stress relaxation is a decrease in stress which occurs under constant strain. Creep will eventually lead to failure, while stress relaxation may result in some permanent deformation, but at some point, the stress level becomes too low and deformation stops.

Creep is generally associated with temperatures which are considered high for a particular material. However, for polymers and plastics, these are temperatures which are often close to the glass transition temperature. Examples of creep at room temperature can be found in contemporary art where artists drape large amounts of fabric, artificial or natural, to obtain the desired effects, see for example [1].

The creep of wood is a phenomenon which many readers may have observed when looking at a fully loaded (historic) bookshelf. One sees that the shelf is bowed slightly downwards under the weight of all of the books. If one were to remove all of the books, the shelf would remain bowed, even after the elastic component of strain was removed. The geometry of the construction including the books generally prevents the shelf from completely collapsing.

This is different if there is a heavy local load on a wood beam as shown in Fig. 5.4. Here one sees a quern-stone, a hand-driven grain milling stone, which was stored in a wooden frame, see Fig. 5.4a. Due to the high local loading, one of the two beams supporting the stone began to deform over time, that is, began to creep. The beam eventually failed, see enlargement of failed area in Fig. 5.4b.

For metals, creep occurs at temperatures higher than approximately 40% of their melting temperature given in absolute temperature, that is in units of Kelvin. Kelvin is the temperature in degrees Celsius + 273.15. Most metals have fairly high melting temperatures, from hundreds to over 1000°C and most conservators will never be confronted with this issue.

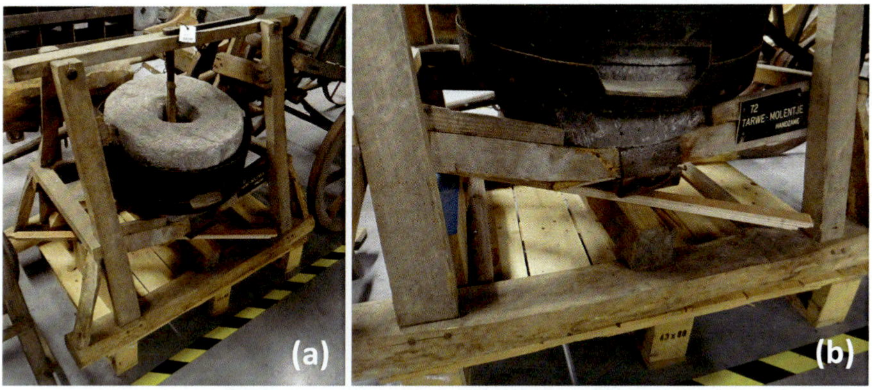

Figure 5.4 Creep failure of a wooden support beam for quern-stone, from the Collectie Bulskampveld, Centrum Agrarische Geschiedenis, Leuven, Belgium. (Photo courtesy: B. Demasure, collection manager) (a) Quern-stone in storage in wooden supporting frame. (b) Creep failure of one of the supporting beams.

There is one notable exception and that has to do with problems with new bass church organ pipes made of pure lead which were made to replace historic pipes [2]. These pipes can be over 4 m long and weigh up to 100 kg. The pipes were found to deform so much at the base after only 20 years (see Fig. 5.5) that they could no longer be tuned. The historic pipes which contained small amounts of impurity elements had never deformed in several hundred years. The deformation of the new lead pipes was found to be due to creep caused by the placement of heating units for the church behind the organ. Pure lead has a melting temperature of about 600 K. 40% of that is 240 K which is approximately –33°C. Historically, the average temperature in a church would be higher than this value, but the addition of heating raised the average temperature near the organ enough that new lead bass pipes began to seriously deform after only 20 years. Furthermore, the lack of impurity elements reduced the resistance of the new pipe materials to creep as compared to the historic pipe materials.

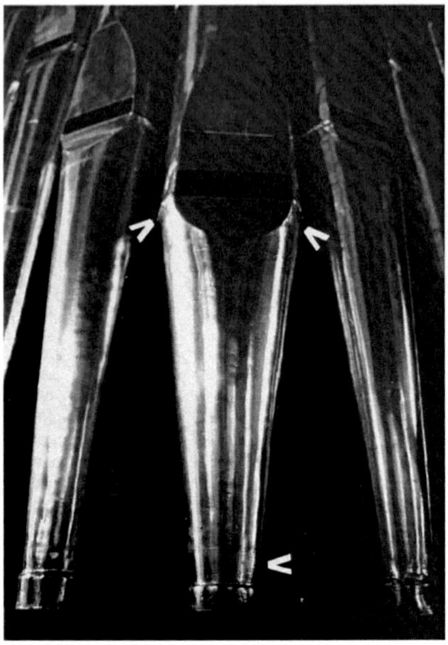

Figure 5.5 Deformation (bulging) at the foot and labium of a bass organ pipe due to creep (white arrows).

5.2 Crack Growth and Fracture Mechanics

5.2.1 Basics

For all of the types of mechanical loads which were discussed in Chapters 2 and 3, the ultimate failure of a material or object was tacitly defined as fracture, meaning that the material or object broke into two pieces. However, the definition of failure in conservation terms is much more conservative, and in fact, in industry, failure is not always defined in terms of the ultimate tensile strength and fracture. We have already mentioned that failure can be defined as excessive plastic deformation, or that a manufacturer can design certain non-critical parts to fail after a few years in order to help increase the sales of a new model of an automobile.

Another form of failure which conservators are confronted with is crack formation and crack growth. Crack formation is a complex phenomenon which even in the engineering world is not completely understood. In conservation practice, it is most often associated with cyclic loads due to long-term cyclic changes in temperature and relative humidity, but also vibrations and cyclic shock loads.

The question often posed by conservators is, however, whether the cracks are stable or will continue to grow. This is, in fact, a question of concern in many applications in the engineering world. If a crack forms in an aircraft or power plant component under cyclic load, it does not mean that the component will fracture immediately. The question is then how long one will accept that the crack grows until it reaches some length when one should consider repairing the component.

The answer to this and related questions is part of the field known in materials science and engineering as fracture mechanics. For the purposes of this book, one can see this field as an extension of the concepts of the strength of materials, σ_{UTS}, and stress concentrations and the stress concentration factor, k_c, discussed in Section 2.6. In the case of a stress concentration, failure, that is, fracture of the material, will appear to occur when the stress in the material reaches a stress much lower than the material's strength. In fact, the stress at a stress concentration such as a small hole or notch has already locally reached the strength of the material.

As we noted at the end of Section 2.6, a crack has a radius at its tip that is so small, that k_c essentially approached infinity. However, when the material at the tip locally reaches the yield stress, plastic deformation occurs which blunts the effect of the sharp radius. Since the concepts of stress and stress concentration cannot deal with a crack, engineers make use of the concept of the so-called stress intensity factor, K, to describe the stress field driving a crack. Note that the stress intensity factor is usually written using a capital letter, K, in order to avoid confusion with the stress concentration factor written with a small letter k_c. Mathematically, the stress intensity factor is written as

$$K = Y \times \sigma \times \sqrt{\pi \times a}, \quad (5.1)$$

where K is the stress intensity factor in units of MPa-m$^{1/2}$, Y is a dimensionless geometric factor, which for rough calculations is often assumed to be 1, σ is the stress in MPa one would calculate if there was no crack (force divided by area) (in the technical literature, it is often called the global or far-field stress), $\pi \approx 3.14$, and a is the length of the crack in mm (see Fig. 5.6 for examples of the definition of a).

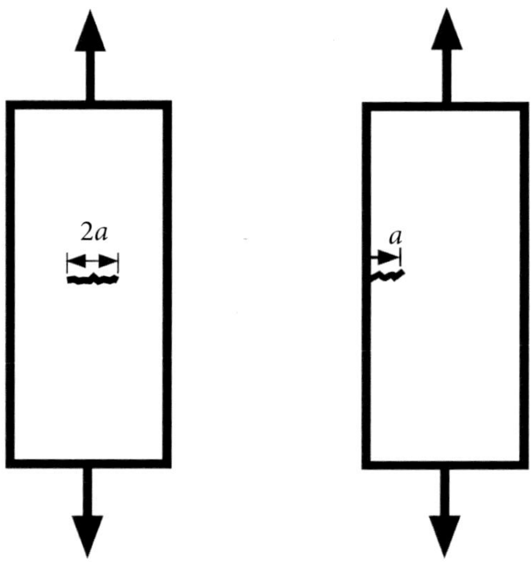

Figure 5.6 Definition of crack length for the calculation of the stress intensity factor, K.

Under static loading, a crack will begin to grow when the combination of the stress, σ, and the crack length, a, results in a stress intensity factor greater than a critical value of K. This value is known as the fracture toughness, K_c. Note that this is not the same as the toughness of a material without a crack as described in Section 2.5. "Begin to grow" is actually a euphemism; the crack will actually shoot through the specimen at the speed of sound as it fails. This assumes that the (testing) environment does not have an effect on the mechanical behavior of the material, see the following section.

As with the case of stress concentrations, the applied global stress when the material fractures will be less than the strength of the material. However, because K is dependent on both this global stress σ, and the crack length, a, any combination of σ and a resulting in a value of K greater than K_c will lead to failure. We would require a higher global stress to fracture a specimen with a short crack than one with a long crack.

There are three basic modes of crack growth, see Fig. 5.7. Mode I crack growth is perpendicular to the load (Fig. 5.7a). Mode II crack growth is in response to a shear load parallel to the crack surface, Fig. 5.7b. Mode III crack growth is in response to a torsional (twisting) type of load, Fig. 5.7c. We then refer to the corresponding stress intensity factors using a subscript I, II or III, that is, K_{I}, K_{II} or K_{III}. The critical stress intensity factors in the three modes are then, K_{IC}, K_{IIC} or K_{IIIC}. Much of the technical and commercial literature refers to mode I loading, and thus one usually finds values of K_{IC} in published tables.

Under cyclic loading, a material or object with a crack is loaded with a cyclic stress intensity factor. Analogous to

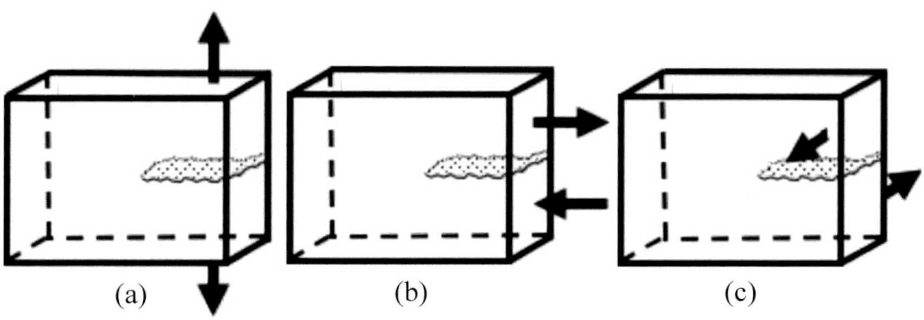

Figure 5.7 Three modes of crack growth. (a) Mode I under tensile loading. (b) Mode II under shear loading. (c) Mode III under torsion loading.

the description of fatigue stresses in Section 3.2, we can describe the cyclic stress intensity factors in terms of either the amplitude, ΔK, or the maximum value of K during a cycle, K_{max}. Crack growth can occur at levels of K_{max} which are less than the fracture toughness, K_c. However, remember from Eq. 5.1 that K is dependent on the applied stress, σ, and the crack length, a. If the global stress amplitude, $\Delta\sigma$, stays constant, the maximum stress intensity, K_{max}, will still increase as the crack grows, that is, as a increases. This means that the crack will grow faster even though the stress amplitude is constant because K_{max} increases as the crack length increases. We say that the crack growth rate, that is, the rate per cycle, $\Delta a/\Delta N$, will increase. In the engineering literature, $\Delta a/\Delta N$ is more commonly written as $\mathrm{d}a/\mathrm{d}N$.

For many materials, the relationship between $\mathrm{d}a/\mathrm{d}N$ and ΔK is written as

$$\frac{\mathrm{d}a}{\mathrm{d}N} = C \times \Delta K^n, \tag{5.2}$$

where $\mathrm{d}a/\mathrm{d}N$ is the crack growth rate in mm/cycle, C and n are experimentally determined constants, and ΔK is the amplitude of the cyclic stress intensity factor.

Substituting equation 5.1 for K, we get

$$\frac{\mathrm{d}a}{\mathrm{d}N} = C \times (Y \times \Delta\sigma \times \sqrt{\pi \times a})^n. \tag{5.3}$$

A typical graph showing the relationship between crack growth rate, $\mathrm{d}a/\mathrm{d}N$, and ΔK is shown in Fig. 5.8. This graph is a double-logarithmic or so-called log–log graph. Each division on both the vertical and the horizontal scales is a factor of ten different from the neighboring divisions. At low values of ΔK, $\mathrm{d}a/\mathrm{d}N$ is very low. In fact, many materials have a threshold, ΔK_{th}, below which there is no crack growth. $\mathrm{d}a/\mathrm{d}N$ then increases linearly on the log–log plot as ΔK increases until the maximum value of K exceeds the fracture toughness, K_C.

Mathematical integration of Eq. 5.2 (or Eq. 5.3 which is a specific form of Eq. 5.2) forms the basis for experimentally based models which allow engineers to predict how many cycles are necessary for a crack to grow from an initial length to some length considered to be critical. It is unlikely that conservators will need to use these directly in their work. However, the fact that the global stress and the crack length both appear in the definition of the stress intensity factor (Eq. 5.1) is useful for understanding of what occurs or will not occur in certain types of objects where cracks are present, as opposed to the situation in Chapters 2 and 3 where they are not.

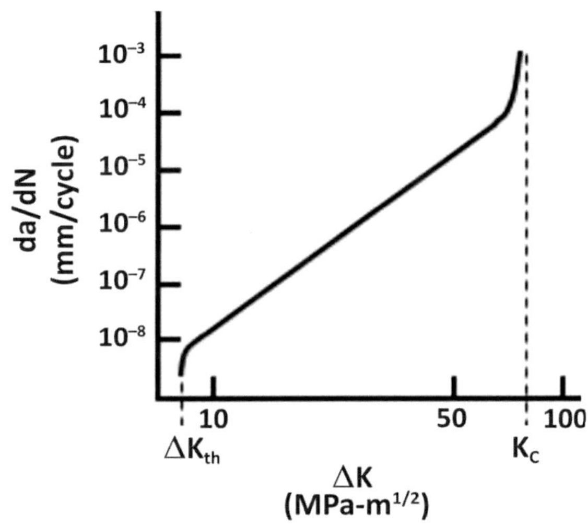

Figure 5.8 Schematic crack growth rate, da/dN versus stress intensity amplitude ΔK diagram. Below the threshold, ΔK_{th}, the crack does not grow. If the maximum stress intensity goes above the fracture toughness, K_C, the material fails.

5.2.2 Craquelure and Fracture Mechanics

A well-known example of cracking in works of art which is often misunderstood in terms of what is happening mechanically is the formation of craquelure and eventual loss of paint during the aging of an oil painting. The craquelure that is visible on a painting is a pattern of cracks which has grown perpendicular to the surface of the painting (see Fig. 5.9a) (and recall Fig. 1.1b). In most cases, it is the result of internal stresses as was described in Section 2.7, combined with the aging of the paints. The internal stresses can be static in nature, for example, due to the drying and aging of the paint, or cyclic, due, for example, to climate changes. The paint layer is stressed in tension in a direction parallel to the surface of the paint so that the mode of cracking is mode I. The cracks grow perpendicularly into the paint film.

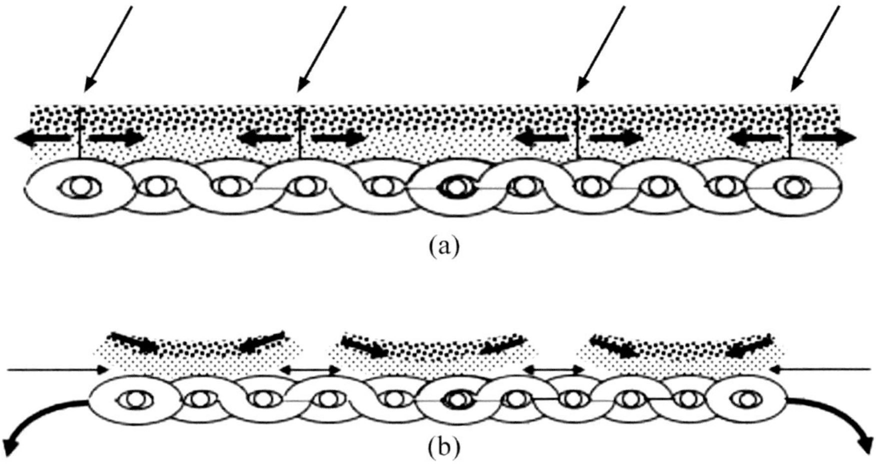

Figure 5.9 Cracking in oil paint layers (a) Formation of craquelure: mode I cracking (thin arrows) perpendicular to the surface due primarily to internal stresses (thick arrows) parallel to the paint surface. (b) Flaking and paint loss due to crack growth in the interface parallel to the surface (thin arrows) due to mixed mode (I, II) internal static stresses and external cyclic bending strains (thick arrows).

Once the craquelure pattern has formed, crack growth into the painting surface ceases. This is because the mode I stresses parallel to the paint surface have been released simply because the paint film is no longer continuous. Furthermore, the cracks have also reached underlying paint layers or the canvas itself, both of which are perpendicular to the crack growth direction. In addition, those materials have different mechanical properties than the paint layers above, which can also blunt further growth in the vertical direction.

Conservators often are concerned that craquelure on older paintings will continue to grow due, for example, to vibrations in transport or poor climate controls. In this sense, this is highly unlikely since the internal mode I stresses have already been released and the cracks have nowhere to go. In terms of Eq. 5.3, a is no longer increasing and $\Delta\sigma$ has essentially gone to zero, so that ΔK is below the threshold ΔK_{th} value for further crack growth.

However, what can happen under (cyclic) bending loads is that the individual paint segments in the craquelure pattern will begin to lift up and eventually fall off. This process, commonly known as "flaking" or "cupping" is a case of crack growth in the interface between the paint and the ground layer or the canvas, that is, parallel to the paint surface (see Fig. 5.9b). This is an example of a more complex, mixed mode of cracking. The formation of craquelure allows the top surface of the paint layer to contract relative to the subsurface layers, responding to the internal residual tensile stresses at the surface formed when the paint was originally drying. This cupping induces bending at the edges of the flakes which is essentially mode I loading. Bending of the canvas due to cyclic loads perpendicular to the canvas can cause further crack growth in the flake/ground or flake/canvas interface as a result of the local bending as the interface opens and closes (local mode I bending strain), but also shear stresses (mode II). This is not the same process as the formation of craquelure. Because of the small dimensions and subsurface nature of the process, this type of crack growth is virtually impossible to observe and monitor.

5.2.3 Wooden Objects and Fracture Mechanics

Fracture mechanics considerations also provide the basis for understanding the (lack of) further crack growth in wooden objects such as panel paintings, furniture, polychrome sculpture and other painted objects. As was discussed for cradled panel paintings in Section 3.4, such cracking is often the result of internal stresses caused by cyclic changes in climate, in particular, humidity. Similar to the discussion for craquelure in oil paintings in the previous section, concerns are often expressed that such cracks will continue to grow. As with paintings, the presence of cracks means that the internal stresses have been released to a great extent. The two halves of the object on either side of the crack are now free to expand or contract. Unless that movement is blocked again, for example, by some kind of repair, a change in climate in terms of humidity or temperature will no longer lead to the buildup of internal stress, σ, needed to continue to drive the crack. The stress intensity factor, K, will remain below the threshold value, K_{th}, and the cracks will no longer grow. In practice, this means that the crack will remain stable in a stable museum environment. On the other hand, care must be taken if the object of, for example, a religious statue, is somehow restored (repaired) and then placed in its original historic location in, for example, an unclimatized church.

5.3 Stress–Environmental Interactions

In the discussion of the basic concepts of the mechanical properties and testing of materials, the tacit assumption was that the environment is not chemically aggressive. We have considered changes in humidity (water vapor) in terms of the internal stresses caused by physical swelling and shrinking, but not in terms of the creation of some new chemical species. We have also briefly looked at the effects of temperature in terms of creep, but not in terms of any acceleration of chemical reactions due to higher temperature.

An aggressive chemical environment combined with stresses can often accelerate crack initiation and crack growth leading to early damage and failure. While this sounds reasonable, there is often a hidden danger, namely, that the effect is a case of "one plus one equals three." In many such cases, the mechanical load itself is not a problem for the object, and the environmental effect by itself is also negligible. However, the combination of the two can lead to serious damage of the object.

The acceleration of crack growth in aggressive environments is known under different names, including stress corrosion cracking (SCC), environmentally assisted cracking (EAC), corrosion fatigue and environmental stress cracking. The first three terms refer to metals and metal alloys, and are common problems in industry under conditions not commonly found in the conservation of cultural heritage except for, perhaps, objects of transport and industrial heritage which may still function to some extent, or some maritime heritage. The SCC of stainless steels is a well-known example in industry. Certain stainless steels are used for their corrosion resistance in chloride–containing environments such as that found near salt water. They can also carry considerable stresses in neutral environments with the presence of cracks as long as the stress intensity factor is less than K_C. However, if a cracked object is stressed in a chloride–containing environment, cracks can grow under static loading even though K is less than K_C, and under cyclic loading they grow faster than the case when cyclic loading is conducted in a non-aggressive environment. Certain bronzes and brasses may also suffer from SCC under certain environmental conditions, among others, heavy bronze objects with high residual stresses in seawater.

In the conservation world, one is more likely to come across cases of environment stress cracking in polymers, which is sometimes also referred to as stress corrosion cracking. In Section 2.7, we introduced the phenomenon of crazing in plastics due to internal stresses. Certain solvents contained in cleaning solutions or adhesives can cause crazing in certain polymers and plastics. Pre-existing crazing can also be made worse since it provides an easy diffusion path for the solvents into the crazed interior of the material.

5.4 Friction and Wear

One of the everyday terms used to describe the aging of works of art and cultural heritage is "wear and tear." Typical examples are the thinning and tearing of historic textiles, the wear of moving parts in works of art in motion (Fig. 5.10a) and the wear of the treads of wooden or even historic or ancient stone staircases (see Fig. 5.10b).

The words "wear" and "tear" are, of course, the results of some kind of mechanical force and that force is referred to as "friction." In mechanical engineering, there is, in fact, an entire field dedicated to these properties, tribology. Tribology is the study of the friction, wear and lubrication of materials. In this section, the concepts of friction and wear as they apply to conservation will be introduced. We will close with a short observation on lubrication.

5.4.1 Friction

Following the order of Chapter 2, which begins with force, we will first begin with the definition of **friction**. Friction is the resistance of two bodies to each other as they move in contact relative to each other. This relative motion can be sliding, such as sliding a box or table across the floor, or rolling, the movement of a bicycle or an automobile on a road. In order to move the two bodies relative to each other, one thus needs a force to overcome the frictional resistance force.

Imagine having to move a wooden crate which is too heavy to pick up. In order to move it, one could try to slide it across the floor. In order to get the crate moving, this will require a force, F, to overcome the friction between the crate and the floor (see the schematic diagram in Fig. 5.11). F can be calculated as

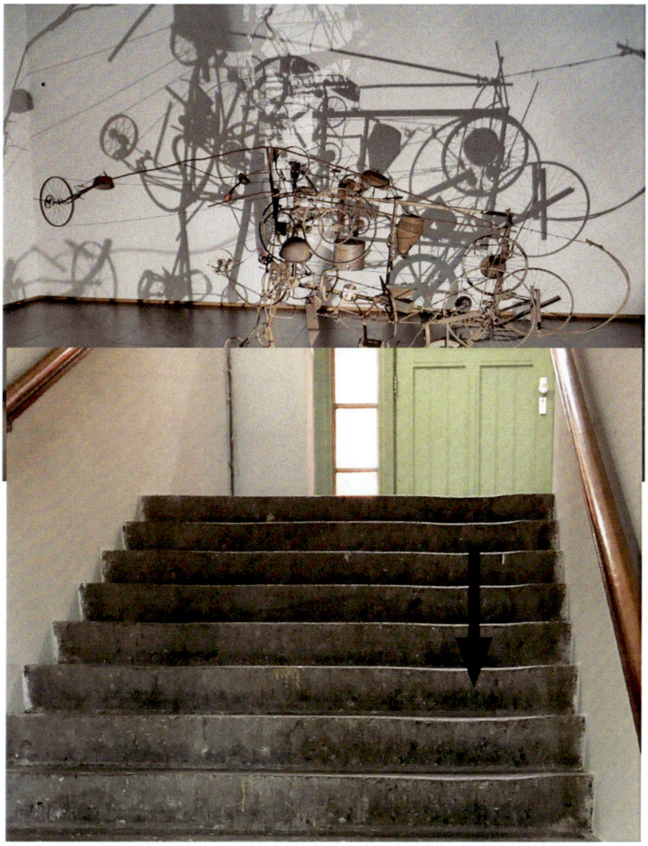

(b)

Figure 5.10 Friction and wear are considerations in the conservation of, for example (a) works of art with moving parts ("Le cyclograveur" (1960) by Jean Tinguely, in a retrospective exhibition (2016) at the Stedelijk Museum Amsterdam, The Netherlands) and (b) wear (arrow) of stone stair treads in a historic building.

$$F = \mu_F \times F_N, \tag{5.1}$$

where μ_F is the so-called static **coefficient of friction** of the crate/floor combination, and F_N is the force normal (perpendicular) to the floor applied by the crate.

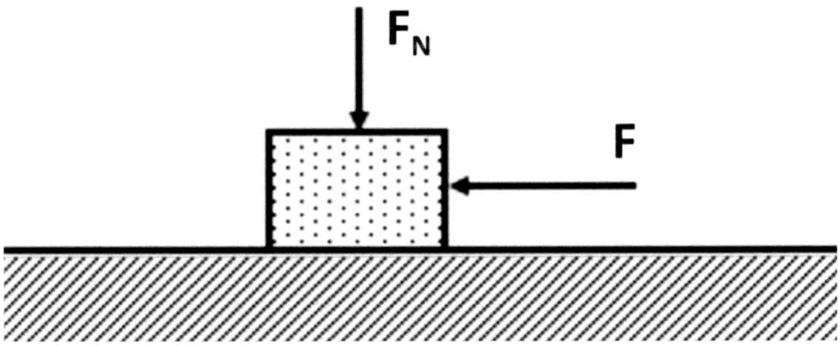

Figure 5.11 Force diagram for the definition of sliding friction. F is the force needed to slide the object along the surface, and F_N is the normal force on the object.

In this case, F_N is the weight (a force) of the crate, $F_N = m \times g$ (recall Eq. 2.3). Note that in this case, the direction normal to the floor happens to be the direction of gravity. However, in general, F_N is defined perpendicular to whatever surface is involved, and that surface can be at any angle as we will see further on in this section.

The higher the value of the static coefficient of friction, μ_F, the more force that will be needed to slide the crate. It is important to note that the static coefficient of friction is a property of the material pair, that is, the crate material and the floor material. The static coefficient of friction is thus NOT the property of a specific material. It is determined by the properties of both surfaces in contact including the surface roughness, the surface chemistry and the environment. These properties determine how the two surfaces may lock into each other, but also how they may attract each other chemically. For the case of the wooden crate on the floor, it would be very easy to move the crate if it were sitting on a surface of ice, for example, ice hockey equipment on the rink. μ_F for wood on ice is approximately 0.02, which means we would only need a force of two percent of the weight of the crate to get it moving in a sliding mode. For a wooden floor, the values of μ_F for wood on wood range between 0.4 and 0.6, which means we need a force which is half the weight of the crate to get it to slide.

Once the wooden crate starts moving, a force is still required to keep it moving, but this is less than it was to get it started. This force is also described by Eq. 5.1, but then the static coefficient of friction is replaced by the so-called dynamic or kinetic coefficient of friction. The kinetic coefficient of friction is generally lower than the static coefficient of friction since the local attraction between the two surfaces has been broken. In practice, most readers have probably had the experience that it is difficult to keep a heavy object in sliding motion because one has to maintain speed and cannot slow down even for a moment. If there is a moment where one weakens, the two surfaces can again "stick" together, and then the higher force is needed to again overcome the static friction to get the object moving again.

Another important point, which for some readers may go against "common sense," is that it does not matter whether one slides the crate shown in Fig. 5.11 in the lengthwise orientation shown or on one of its ends. In other words, for a particular object, the forces are independent of the contact area. In general terms, this is because, even though the pressure is higher for the smaller contact area (force divided by smaller contact area), the effects of surface properties are also less because there is less contact area. (This argument does not, of course, consider the possibility that the object may tip over.)

The concept of friction can come up in a number of cases related to conservation. One example is the display of textiles at an angle when affixing them to a three-dimensional/vertical display structure is not desirable. The question is at what angle, θ, the object will begin to slide down the display structure, for example a flat transparent plate as shown schematically in Fig. 5.12. In such a case, the normal force, F_N, is reduced by a factor of the cosine of the tilt angle, θ, and is written as

$$F_N = m \times g \times \cos\theta. \tag{5.2}$$

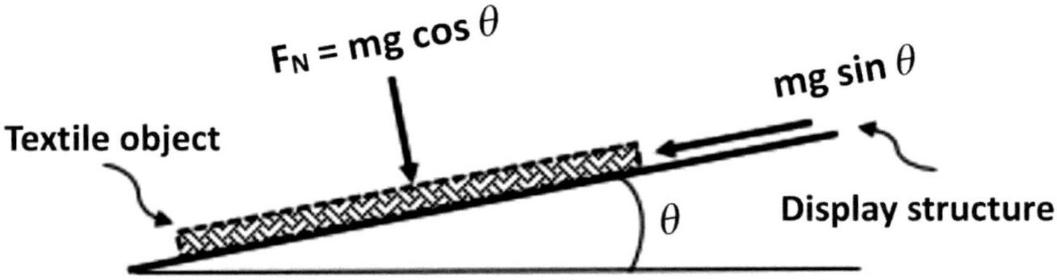

Figure 5.12 Forces acting on a textile object displayed without fixation at an angle.

Recall that the definition of the "normal" force is the force perpendicular to the flat plate, not the direction of gravity. The friction force parallel to the flat plate we need to overcome in order to get the textile to start sliding is then

$$F = \mu_F \times m \times g \times \cos\theta, \tag{5.3}$$

where μ_F is the static coefficient of friction between the textile and the flat plate material.

However, because the object is tilted, there is already a force working to get the textile object to slide. That force is that fraction of the weight of the tilted crate parallel to the flat plate, $m \times g \times \sin\theta$. Recall from geometry that the cosine of an angle decreases as the angle increases, and the sine of an angle increases as the angle increases. Thus, as we tilt the plate towards higher angles, θ, the force needed to get the textile to slide will decrease as you would expect, and the fraction of weight contributing to the force already present will increase. The angle at which these two are equal is the critical angle beyond which the textile will slide. Mathematically, that is the angle, θ_c, at which the two forces become equal:

$$(\mu_F \times m \times g \times \cos\theta_c) = m \times g \times \sin\theta_c \tag{5.4}$$

We can solve for the angle by noting that $m \times g$ sits on both sides of the equation and then rewrite it as

$$\mu_F \times \cos\theta_c = \sin\theta_c$$

followed by

$$\mu_F = \frac{\sin\theta_c}{\cos\theta_c},$$

which is

$$\mu_F = \tan\theta_c. \tag{5.5}$$

The critical angle, θ_c, we are looking for is then the cotangent of the coefficient of friction,

$$\theta_c = \tan^{-1}(\mu_F). \tag{5.6}$$

The cotangent (cot, cotan, $\tan^{-1}$) function can be found on most calculators with basic scientific functions.

5.4.2 Wear

The deformation and eventual damage resulting from the relative motion of two objects in contact is known as **wear**. Wear is generally measured in terms of loss of weight of material and/or a change in surface roughness. Wear can occur in just one relative movement of two objects in contact. However, it is mostly associated with cyclic loading and is thus a cumulative process.

The prediction of how a material wears in contact with another and how fast it wears is still a major field of research in tribology. In general, one can say that softer materials will suffer more wear when in contact with harder materials, and high wear rates are associated with higher normal loads and higher coefficients of friction. However, this is not always the case. There are international standards for friction and wear testing, but when it comes to the actual application, practical simulation testing is often necessary to determine how and how fast a material will wear in contact with another material. For the conservation field, this will also be the case.

There are several basic types of wear described in the tribological literature, four of which conservators are more likely to encounter in their work than the others. The type of wear probably most familiar to conservators and conservation scientists, though perhaps not under this term, is known as **abrasive wear**. This type of wear occurs when hard particles slide across or are rolled into a surface causing scratching and/or removal of material. On a microscale, this is what happens when a conservator or conservation scientist polishes a specimen with micron or sub-micron diamond or alumina particles for cross-section analysis. Dry cleaning of paintings can also lead to abrasive wear on a microscopic scale. Avoiding abrasive wear is one of the main reasons that clean, soft brushes are often recommended for cleaning works of art.

On a macroscale, the wear of stair treads over the years (recall Fig. 5.10b) or for stone stairs over centuries, is also due to abrasive wear. While leather shoes or bare feet are softer than most stair materials, they will always carry bits of sand and dirt which are much harder and can scratch the surface in the same way as polishing materials. Abrasive wear is also a common cause of the degradation of historic carpets in historic buildings which are still in use. Here again, footware carries hard particles which then abrade the nap of the carpet, causing what often is mistaken for the fading of the colors in the carpets, see for example, Fig. 5.13.

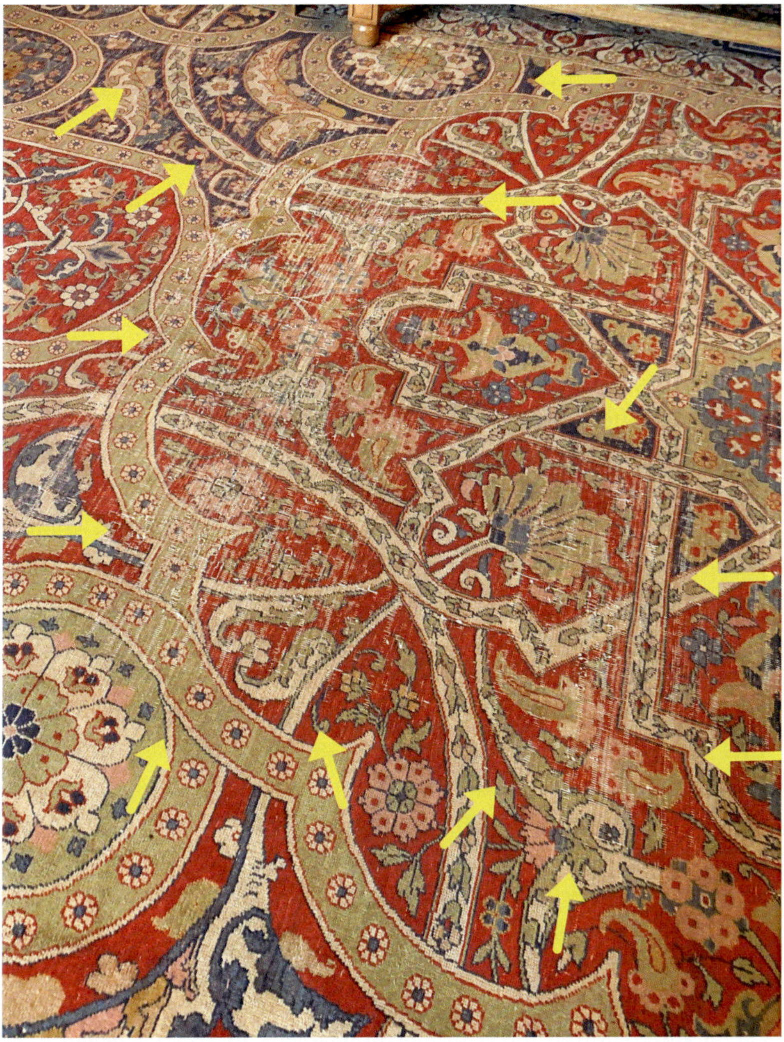

Figure 5.13 Wear of a historic carpet which had been mistaken for color fading, see arrows.

A type of wear related to abrasive wear is **erosive wear** (see Fig. 5.14a). Erosive wear occurs when particles strike a surface at high speed and remove material. The strength of erosion depends on the material and the angle of the stream of particles to the surface. A rule of thumb is that brittle materials such as ceramics sustain higher rates of damage when they are struck perpendicular to the surface, while more ductile materials sustain higher rates of damage when they are struck at angles of 20–30°.

In conservation, erosive wear may be useful for the cleaning of object surfaces. Experiments have been conducted using streams of dry ice or even soft materials such as peanut shells to remove contaminants and corrosion from, among others, metal objects (see, for example, [3]). On the other hand, erosive wear can be an undesirable side effect for simple treatments such as the use of compressed air or streams of water or other chemical solutions to remove dust from objects. Dust often contains hard microparticles of alumina or silicate soil particles which can damage a surface if the air or fluid pressure used is too high. For some types of objects such as those made of PMMA, the use of dust- and oil-free compressed air sources is recommended for this reason.

Another type of wear which conservators may come across in moving parts in contact is known as **adhesive wear** (see schematic diagram in Fig. 5.14b). As the term implies, this type of wear occurs when the two surfaces in contact tend to stick to one another which is often the case when two of the same material are used in contact with each other. In general, the softer material will delaminate, that is, be peeled away by the other material, or the surface material will literally be smeared. This type of wear can be found, for example, in the gear teeth of old clocks where the lubrication has aged and failed, or there was no lubrication at all.

A final type of wear which may occur in special cases is so-called **fretting** wear. Fretting itself is the rapid cyclic sliding contact of two surfaces, most often caused by vibrations (see Fig. 5.14c). The word itself comes from the traditional picture of what parents often do with their hands when they are extremely worried about a child. Objects which may be vibrating/resonating on a hard surface may show signs of fretting wear if the vibration levels are high and continue for long periods of time.

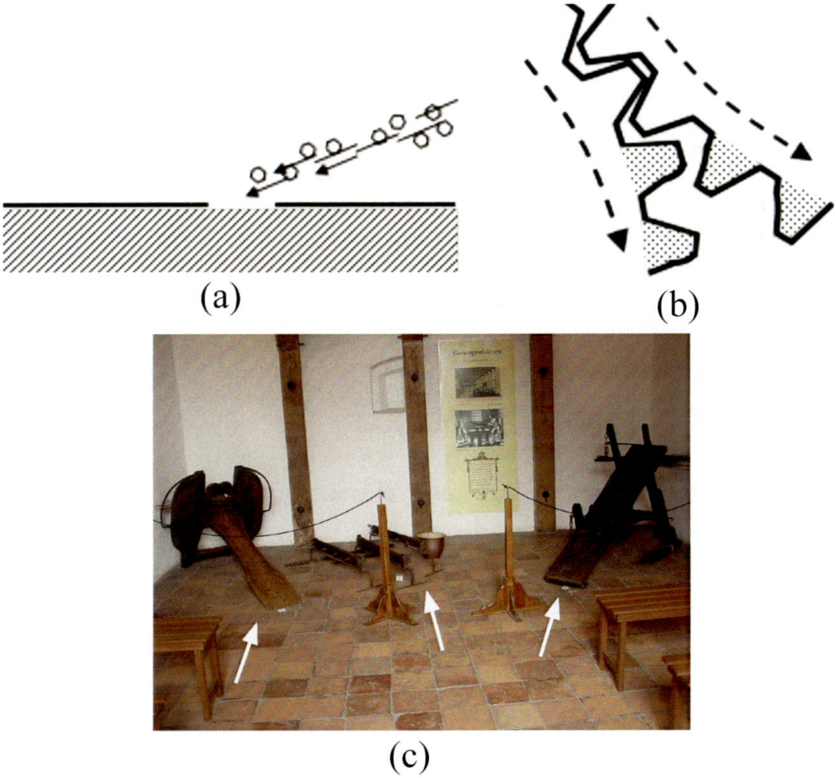

Figure 5.14 Three types of wear which conservators may encounter in their work. (a) Erosive wear due to streams of hard particles. (b) Adhesive wear in gear teeth (schematic, lower two teeth of both gears). (c) Possible locations of fretting wear due to vibration of the objects in contact with the ceramic floor during construction.

5.4.3 Lubrication

A detailed discussion of lubrication is beyond the scope of this book. However, it is very important for conservators to know that one does not just use any lubricant during the conservation/repair of moving parts in contact with other moving or non-moving parts. Most modern lubricants are designed for specific applications, that is, loads, environments, and the specific materials of which an object is made.

The makers of a historic object may or may not have known that they should use a proper lubricant. The principles of conservation ethics urge conservators to use original materials where possible. However, given the current state of the art in tribology, conservators should seriously think about how far they want to go with that principle, if using an original lubricant will lead to (repeated) irreparable chemical and/or physical damage in the future.

5.5 Life Prediction for Vibrations and Other Complex Cyclic Loads

In Chapter 3, the concepts of cyclic loading and life prediction were introduced. The discussion was based on a simple waveform for a cyclic load (Fig. 3.3) and the concept of the *S–N* or Wöhler diagram for fatigue life prediction (Fig. 3.6). In order to determine the fatigue life of a material or object, one simply drew a horizontal line from the maximum stress level or stress amplitude to the experimental fatigue curve. By moving vertically down to the horizontal axis, one could then determine the fatigue life of the material or object in number of cycles, N_f.

We noted in Section 3.7, however, that in many situations, a collection or object can be exposed to mixed shock and vibration loads. This leads to the question as to how we can deal with much more common real-life situations where cyclic levels are essentially random. Construction work, rock concerts and transport certainly do not occur at a single stress amplitude and frequency.

A number of highly sophisticated life prediction models have been developed in the engineering world over the last 40 to 50 years to help predict the life of components and products which undergo random cyclic loads. The most important of these are those used in the aircraft industry. For the purposes of this book and for most of your work in preventive conservation, it suffices, however, to introduce the original basis for these models, the so-called Palmgren–Miner rule.

Instead of a simple cyclic load at a single stress amplitude level, imagine now that we have a case where an object is exposed to three different maximum cyclic stress levels, σ_1, σ_2 and σ_3 during some event such as a museum loan or a rock concert near a museum. The question is, how much of its fatigue life has it spent, or how much is left?

We therefore need to consider our simple *S*–*N* diagram with all three maximum stress or stress amplitude levels as shown in Fig. 5.15. The Palmgren–Miner rule was developed to predict the cyclic life of an object/material in such situations.

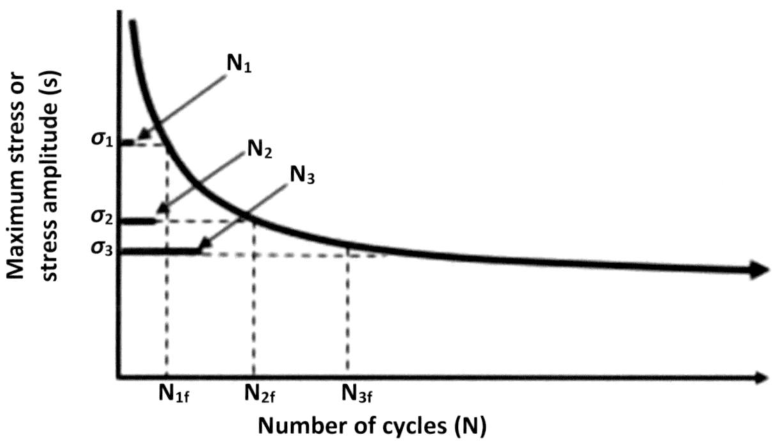

Figure 5.15 Wöhler diagram marked up for calculating the used and remaining life of an object exposed to three maximum cyclic stress levels during one event, see main text.

Instead of thinking in terms of absolute numbers of vibrations or events, the Palmgren–Miner rule has us look at the fraction of life spent at the different cyclic stress levels. If an object were continuously subjected to vibrations at a constant level, σ_1, it would eventually "fail" at a given number of cycles (or time), N_{1f}, and have "served its whole life" at that level. We will give the entire life a value of 1. If the object were only exposed to a constant lower level such as σ_2 or σ_3, it would eventually "fail" at a larger number of cycles, N_{2f} or N_{3f}, respectively, its entire life at those levels, each with a value of 1. The Palmgren–Miner rule states that if an object is exposed to different levels of vibrations, it will fail when the sum of the fractions of life it uses up at each level reaches a value of 1.

To put this a little more mathematically, imagine that an object is exposed to an event with the three different maximum cyclic stress levels in Fig. 5.15, σ_1, σ_2, and σ_3. It spends N_1 cycles at level σ_1, N_2 cycles at level σ_2, and N_3 cycles at level 3. Then

- at level σ_1, it has used up the fraction N_1/N_{1f} of its total fatigue life;
- at level σ_2, it has used up the fraction N_2/N_{2f} of its total fatigue life;
- at level σ_3, it has used up the fraction N_3/N_{3f} of its total fatigue life.

The object will have thus been exposed to $N_1/N_{1f} + N_2/N_{2f} + N_3/N_{3f}$ of its life. If the sum $N_1/N_{1f} + N_2/N_{2f} + N_3/N_{3f}$ is less than 1, then the object will not yet have failed. If the sum $N_1/N_{1f} + N_2/N_{2f} + N_3/N_{3f}$ is equal to or greater than 1, then failure occurs.

In engineering books, you will find the Palmgren–Miner criterion for failure due to mixed cyclic loads written in the more general terms:

$$\sum_{1}^{i} \frac{N_i}{N_{if}} \geq 1, \tag{5.7}$$

which translated is what we just wrote above, that is, if the sum (Greek capital letter sigma, Σ) of the fractions of life at each level, N_i/N_{if} (where i = 1, 2, 3, ..., to however many stress levels there were) is equal to or greater than 1, the object will fail.

Practice exercise 5.1

We will now illustrate the Palmgren–Miner rule, using real numbers. Refer again to Fig. 5.15. Assume the following:

- The total life at the maximum stress level, σ_1, is $N_{1f} = 1000$ cycles.
- The total life at the maximum stress level, σ_2, is $N_{2f} = 10{,}000$ cycles.
- The total life at the maximum stress level, σ_3, is $N_{3f} = 100{,}000$ cycles.

Our object has been exposed to all three cyclic stress levels during one event as follows:

- 100 cycles at level σ_1
- 3000 cycles at level σ_2
- 13,000 cycles at level σ_3.

The fraction of life used at level σ_1 was thus $N_1/N_{1f} = 100/1000 = 0.1$; the fraction of life used at level σ_2 was thus $N_2/N_{2f} = 3000/10000 = 0.3$; and the fraction of life used at level σ_3 was thus $N_3/N_{3f} = 13{,}000/100{,}000 = 0.13$.

The total fraction of life used during this event was then

$$N_1/N_{1f} + N_2/N_{2f} + N_3/N_{3f} = 0.1 + 0.3 + 0.13 = 0.53.$$

The object was exposed to around half of its fatigue life. Note that this means that there is just a little less than half of the total fatigue life available, 1 – 0.53 = 0.47. You can use this calculation to determine if a following event or following events are allowable. If you know the number of cycles the object will be exposed to at the three levels in a future event, you can calculate what further fraction of life would be used up. If that is less than 0.47, you can permit that event to go through.

Given the uniqueness of objects of art and the complexity of the cyclic loads it may be exposed to, it would be a virtually impossible task to literally and quantitatively apply the Palmgren–Miner rule to every conservation situation. However, it is a useful concept when it comes to negotiating allowable levels of cyclic loads, for example, vibrations during construction or during a rock concert (see, for example, Fig. 5.16). Figure 5.16 is a graph of a typical vibration signal plotted as strength (amplitude) in velocity (mm/s) versus time. For the purposes of this discussion, imagine that it is known from previous conservative experience, that amplitude *A* is known to be a limit where no damage has ever been observed for your type of collection. This means that we find ourselves somewhere towards the right side of the *S*–*N* diagram, closer to the fatigue limit. Considering the Palmgren–Miner concept, the percentage of life for the collection used at this level will probably be very low. Amplitude *A* is thus a very conservative value driven by the desire for "absolutely no risk," which in fact, does not exist. Furthermore, by using this limit for the event, the alarm will be triggered quite often and becomes an irritating and expensive exercise for all parties involved.

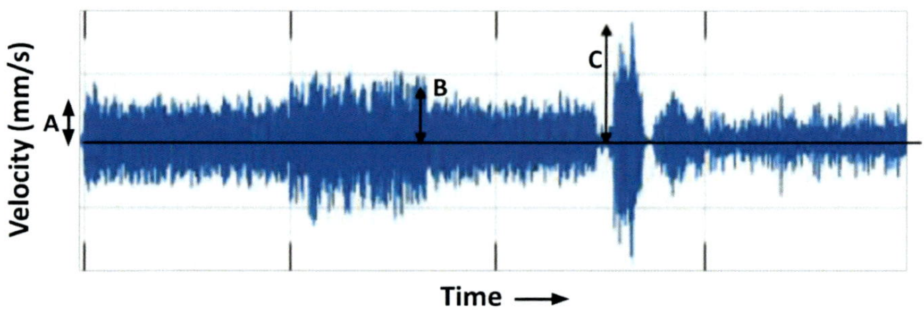

Figure 5.16 Typical vibration signal in terms of velocity versus time. The Palmgren–Miner rule shows that some flexibility is possible when negotiating limits for vibration loads.

Given the idea that the percentage of life used at level *A* is low, the Palmgren–Miner concept implies that one has some negotiating room to raise the trigger level, say to level *B*. If the whole event took place at level *B*, a larger fraction of the life of the collection would be used up than at level *A*. However, in this case, level *B* vibrations make up only around 15% of the entire event, thus the fraction of life used at level *B* is lowered concomitantly. Of course, a risk analysis would have to be conducted to determine how much negotiating room one would want to accept. But by setting the vibration limit level a bit higher for the duration of this event, one would save a lot in irritation and costs, probably at low risk for the collection. In either the case of a limit of levels *A* or *B*, vibration events at level C would not be acceptable.

In conclusion to this section, the Palmgren–Miner rule has been presented to provide the reader an idea of how engineers predict product lives under complex cyclic loads. As a quantitative model for preventive conservation, it has little use, since there is very little material and objects data, read, there is very little *S–N* data to use in calculations. However, the rule does show that there is some room for negotiation when it comes to dealing with vibration sources, since the conservation world has, up to this time, always worked with very conservative limits on cyclic loads where no risk has been accepted.

5.6 Computer Modelling and Finite Element Analysis

Many mechanical issues in conservation can be dealt with a good understanding and use of the basic concepts and equations described for static and dynamic loads in Chapters 2 and 3. These concepts and equations make up what are called **analytical** models, which simply put, are models which we could solve on the "back of an envelope" or using a calculator. However, we have seen that as objects and mechanical loads become more complex, the mathematics of predicting materials strength and life also become more complex. At some point, computational models may be necessary.

The basis for such computational models is the so-called **finite element method (FEM)** or finite element analysis (FEA). In such an analysis, a model of the object is first created in a computer program. In this model, the object is divided up into a mesh of very small elements (see Fig. 5.17). The computer program is designed to calculate how each element reacts to the applied stresses and/or strains. The totality of these reactions determines how the object as a whole reacts to the applied stresses/strains. This macroscopic reaction is based on the fact that the elements are connected to each other, so that two neighboring elements must, of course, react in the same way at their common boundary.

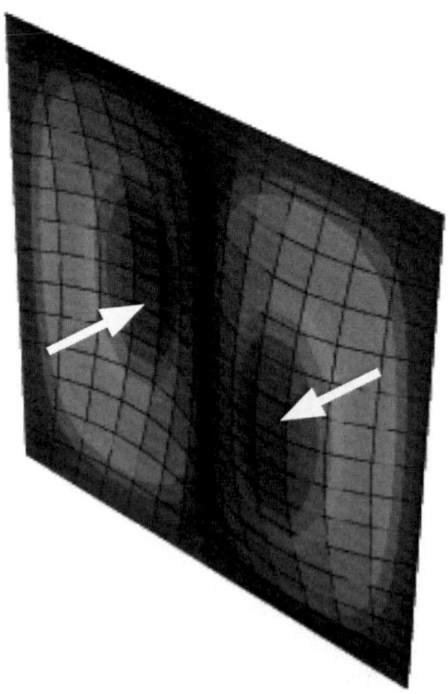

Figure 5.17 Visualization of the finite element analysis of the vibration of a painting canvas. The arrows show maxima in deformation of a secondary resonance mode.

The more elements there are, the better, but the number of elements is limited by the amount of computer time required for the calculations. The more elements used, the more expensive the calculation becomes. The number of elements is thus finite, hence the term finite element analysis. In order to obtain an optimal model, the mesh is designed so that it has many elements at locations which are considered to be critical (fine mesh), and fewer elements where large local stresses or strains are not expected.

The input into such a computational (FEM) model includes analytical materials models ranging from the simple Hooke's law (Eq. 2.10) to more complex models such as the Palmgren–Miner rule we just discussed in the previous section. Engineers who develop such FEM models often have to make certain assumptions about how the analytical models are applied and what material properties should be used if data are scarce, and conduct experiments to verify the computed results. When considering the results of such modelling, the reader must always ask if the results make sense, especially since publications reporting the results tend to be written in pure technical language with fancy colored 3D graphs. In the worst case, the adage, "garbage in, garbage out" applies.

An example of the visualization of the results of a finite element analysis of the vibration of painting canvas material is shown in Fig. 5.17. One sees the deformation of a canvas due to vibrations at a secondary resonance frequency. The model was built to predict the frequencies at which a canvas of a specific size and material would resonate [4]. One of the main assumptions was that the weave of the canvas could be modelled as a thin plate. This involved a separate FEM model of the weave involving, among others, the curvature of the threads in the weave, the weave structure and density, and friction between threads as the canvas is mechanically loaded. The mechanical properties of the canvas material were determined experimentally, and the *E*-modulus for the two directions of the canvas was used as input, assuming that

Hooke's law for the elastic behavior of the material applies. One of the difficulties in the development of this model was determining the value of the E-modulus to be used based on the stress–strain diagram for textiles, recall the schematic diagram in Fig. 2.10.

References

1. Van Aubel, C., Wei, W., Scholte, T., de Groot, S., van Keulen, H., and de Groot, E. (2017). Stretch in Space. Research into the mechanical perperties of installations of knitted polyamide under stress, *Proc. Future Talks 015 – Processes. The Making of Design and Modern Art. Materials, Technologies and Conservation Strategies*, ed. Bechtold, T., October 28–30, 2015, (Die Neue Sammlung–The Design Museum, Munich, Germany) pp. 29–38.
2. Wei, W., and de Groot, M. (2001). Investigation into the causes for the sagging of lead and lead-tin organ pipes, *ISO Journal*, 10, pp. 60–68.
3. Van der Molen, R., Joosten, I., Beentjes, T. and Megens, L. (2011). Dry-ice blasting for the conservation cleaning of metals. *Proc. Metal 2010: Interim meeting of the ICOM-CC Metals Working Group*, October 11–15, 2010, Charleston, South Carolina, USA (ICOM-CC, Paris).
4. Chiriboga-Arroyo, P.G. (2013). Finite element modelling of vibrations in canvas paintings, PhD dissertation, Technical University of Delft, Faculty of Aerospace Engineering.

Epilogue

This book was designed to provide conservators and other cultural heritage professionals, with little or no technical/ engineering background, a basic introduction to mechanical properties and testing. At a minimum, it is hoped that the reader has come away with a good understanding of the basic terminology used in this engineering field known generally as mechanics of materials. A good understanding of the concepts behind these technical terms is necessary for properly designing mechanical testing and specimens for applications in conservation practice. In addition, proper use and understanding of this terminology is necessary to effectively communicate with testing engineers from outside the conservation field, and to be able to more critically assess advanced testing and computer modelling methods which may be introduced in conservation science.

Although a large number of terms and concepts have been introduced in this book, it is most important that the reader understands the following concepts:

- The difference between force (F) and stress (σ)
- The difference between deformation/elongation (l) and strain (ε)
- The relationship between stress and strain and the concept of material stiffness (E)
- The difference between material stiffness (E) and the geometric stiffness of an object
- The difference between static, shock and cyclic loads
- The concept of cumulative damage due to cyclic loads such as vibrations or internal stresses due to cyclic changes in temperature or humidity

Further, it is hoped that the reader has obtained a basic understanding of how to address questions of mechanical properties which they may face in their daily practice. A number of exercises including calculations and simple mechanical tests were included in this book to give the reader the opportunity to practice using the basic concepts discussed in the various chapters. These calculations and simple tests were designed based on real-life situations. For conservation practices with no access to mechanical testing equipment, they could, to a certain extent, be used in the conservation studio to make first rough estimates of mechanical properties and to help with materials selection.

The user is, however, cautioned that the exercises were adapted to the teaching goals of this book. As discussed in Chapter 5, many conservation issues involve mechanical loads which are more complex than those described in the book. Referring to Practical exercise 2.1, for example, glues as well as adhesives are not loaded in pure shear, but also have a component of peeling. Shear and peel testing are more difficult to carry out and interpret, as was discussed in Section 4.7. For a rough comparison of glues for larger objects such as historical furniture, the calculation in Practical exercise 2.1 might still be sufficient. However, for answering questions related to, for example, the effect of environmental changes on paint flaking and cupping, or the selection of hinges for hanging works of art on paper, more careful consideration of the testing methods is required, eventually in consultation with a specialist in engineering mechanics and testing.

It is hoped that this book has provided the encouragement and motivation for readers to more actively undertake proper testing in order to solve questions about mechanical properties. In the long term, this will be a valuable contribution to solving yet another major problem in the area of mechanical testing in conservation: the scarcity of useful mechanical properties data due to the lack of attention given to such testing in past conservation curricula and practice. This book can be considered a success if even a small step is taken in that direction.

Appendix

Conversion of Selected English (Imperial) and SI (International) Units Used in This Book

Force

1 pound-force (lb_f) = 4.45 newtons (N)

To get N from lb_f, multiply the number of lb_f by 4.45.
To get lb_f from N, divide the number of N by 4.45.

Mass

1 pound (lb_m) $\cong$ 0.45 kilograms (kg)

To get kg from lb_m, multiply the number of lb_m by 0.45.
To get lb_m from kg, divide the number of kg by 0.45.

Stress

1 pound per square inch (psi) = 0.00689 mega-Pascal (MPa)
(Recall that 1 MPa is the same as 1 N/mm^2.)

To get MPa from psi, multiply the number of psi by 0.00689.
To get psi from MPa, divide the number of MPa by 0.00689.

Length

1 inch (in) = 25.4 millimeters (mm)

To get mm from in, multiply the number of in by 25.4.
To get in from mm, divide the number of mm by 25.4.

Area

1 square inch (in^2) = 645.2 square millimeters (mm^2)

To get mm^2 from in^2, multiply the number of in^2 by 645.2.
To get in^2 from mm^2, divide the number of mm^2 by 645.2.

Area of a circle

$A = \pi r^2$
$\pi \approx 3.14$
r = radius of a circle

Volume of liquids

1 fluid ounce (fl. oz) = 29.57 milliliters (ml or mL)

To get mL from fl. oz, multiply the number of fl. oz by 29.57.
To get fl. oz from mL, divide the number of mL by 29.57.

Velocity

1 inch per second (in/s) = 25.4 millimeters per second (mm/s)

To get mm/s from in/s, multiply the number of in/s by 25.4.
To get in/s from mm/s, divide the number of mm/s by 25.4.

Index

For Product Safety Concerns and Information please contact our EU representative GPSR@taylorandfrancis.com Taylor & Francis Verlag GmbH, Kaufingerstraße 24, 80331 München, Germany